GEOLOGICAL GUIDE TO HALEAKALA NATIONAL PARK

BY

RICHARD ROBINSON

Cover photo due to the extreme generosity of the outstanding nature photographer QT Luong. Winner of six national book awards and author of: Treasured Lands: A Photographic Odyssey through America's National Parks

The New York Times Book Review:
"No one has captured the vast beauty of America's landscape as comprehensively."

TABLE OF CONTENTS

ACKNOWLEDGEMENTS

The author wishes to thank Sally J. Bensusen, Sr. Graphic Designer NASA Science Program Support Office, Global Science and Technology, Inc. for providing me with links to the satellite images used in this book. These images *are not* photographs, but images created from data acquired by a satellite and made into "visualizations" by the various imaging teams. All images have been used via the courtesy of NASA".

In addition, Simon Balm's computer assistance was indispensable. His help in dealing with Kindle Publishing made this book possible. Also, Brian Pena has been a lifesaver solving many problems on many occasions! This book could not have been made without their assistance.

FORWARD

- This book, written for the *general public*, will give you mile by mile descriptions of the geology from Kahului to Haleakala's summit and from Hana to the Kipahulu Area of Haleakala National Park.
- In some areas, the geology is very simple, and a brief description will be given at the indicated mileage points.
- In other areas, the geology is more complex, and the main highlights will be given, followed by, in a few cases, optional details.

ODOMETERS

- Try to set your odometer as close as possible to the starting point given in the road logs.
- However, all odometers are not created equal. Odometer deviations of 0.1-0.3 miles are not uncommon.
- However, these minor differences should not cause you any serious problems.

MILEAGE DEVIATIONS

- Any deviations from the road log mileages must be added to those given in the road log.

DISCLAIMER

- **Use caution and common sense while using these geologic road logs.**
- **Drive, park, and hike safely.**
- **Do not go any place you may feel is unsafe.**
- **Do not venture away from the beaten path.**
- **After you park your car, do not leave valuables in plain sight, even if the car is locked.**
- **The author and publisher assume no liability for accidents, injury, or any losses or deaths by individuals or groups using this publication.**
- **The author of this geology guide assumes no responsibility for any accidents, injuries, or deaths based on these descriptions.**
- ***Users of this book must assume full responsibility for their actions.***

ABOUT THE AUTHOR

- The author of this book taught geology at a southern California college for 41 years.
- Many trips were taken to Maui to make these road log mileage points as accurate as possible.

ADDITIONAL BOOKS BY THIS AUTHOR

- Geological Guide to Oahu (color version)
- Geology of Oahu (black and white version)
- Illustrated Guide to the Island of Hawaii (color version)
- Island of Hawaii Geological Guide (Black and white version)
- Geological Guide to Hawaii Volcanoes National Park (6x9 color version)
- A Guide to the Geology and Geography of Kauai (color version)
- Geology and Geology of Kauai (black and white version)
- *All of the above books are available on Amazon.*
- *All prices set by Amazon.*

IMPORTANT TRAVEL INFORMATION

- Before beginning these road logs, be sure to *start with a full tank of gas!*
- Also, even if you have a GPS unit, be sure to also have a *good* road map. GPS units are *not* infallible!
- In the event of an emergency, be sure your cell phone is charged.
- However, in some rural areas you may not have cell phone service.
- Loss of cell phone service could be especially true on the south side of Haleakala.
- Also, have your roadside assistance phone numbers available should the need arise.
- These phone numbers are generally on your rental car's key ring.
- Above all, watch for careless drivers and be sure that you are not one yourself.
- At times, to prevent a particular mileage point from catching you "off guard", some of the mileages to a particular location have been shorten by 0.1 miles.

HAWAIIAN ISLAND DIRECTIONS

- Note: you will hear locals using *mauka* and *makai* when giving directions.
- Islanders do not use the terms inland and seaward.
- *Mauka* is used for *inland*.
- *Makai* is used for *seaward.*

The following websites will allow you to speak like an islander

To *hear* Maui *place names* pronounced, go to:

http://hawaiian-words.com/hawaii-place-names

To *hear* Hawaiian *words* pronounced, go to:

http://hawaiian-words.com/common

.....

6

GENERALIZED MAP OF TOWNS AND ROADS ON MAUI

Modified from Maui Hotels.com

ESTIMATED DRIVING TIMES ON MAUI (without traffic)

Kahului to Lahaina: 40 minutes
Kahului to Kaanapali: 45 minutes
Kahului to Kapalua: 55 minutes
Kahului to Maalaea: 15 minute
Kahului to Hana: 2 hours and 40 minutes
Kahului to Kihei: 15 minutes
Kahului to Wailea: 30 minutes
Kahului to Makena: 35 minutes
Kahului to the summit of Mount Haleakala:
        1 hour and 30 minutes
Kahului to Iao Valley State Park: 15 minutes
Kaanapali to Lahaina: 5 minutes
Kaanapali to Kapalua: 10 minutes
Kaanapali to Maalaea: 30 minutes
Kaanapali to Hana: 3 hours and 25 minutes
Kaanapali to Kihei: 40 minutes
Kaanapali to Wailea: 55 minutes
Kaanapali to Makena: 1 hour
Kaanapali to the summit of Mount Haleakala:
        2 hours and 15 minutes
Kaanapali to Kahului: 45 minutes
Kaanapali to Iao Valley State Park: 45 minutes

Kihei to Lahaina: 35 minutes
Kihei to Kaanapali: 40 minutes
Kihei to Maalaea: 10 minutes
Kihei to Hana: 2 hours and 55 minutes
Kihei to Wailea: 15 minutes
Kihei to Makena: 20 minutes
Kihei to the summit of Mount Haleakala: 1 hour
        and 45 minutes
Kihei to Kahului: 15 minutes
Kihei to Iao Valley State Park: 25 minutes

Wailea to Lahaina: 50 minutes
Wailea to Kaanapali: 55 minutes
Wailea to Kapalua: 1 hour and 5 minutes
Wailea to Maalaea: 25 minutes

Wailea to the Hana Lava Tube (a.k.a. Kaeleku
        Caverns): 3 hours
Wailea to Hana: 3 hours and 10 minutes
Wailea to Kihei: 15 minutes
Wailea to Makena: 5 minutes
Wailea to the summit of Mount Haleakala:
        2 hours
Wailea to Kahului: 30 minutes
Wailea to Iao Valley State Park: 40 minutes
Wailea to Lahaina: 50 minutes
Wailea to Kaanapali: 55 minutes
Wailea to Kapalua: 1 hour and 5 minutes

Kapalua to Lahaina: 15 minutes
Kapalua to Kaanapali: 10 minutes
Kapalua to Maalaea: 40 minutes
Kapalua to Hana: 3 hours and 35 minutes
Kapalua to Kihei: 50 minutes
Kapalua to Wailea: 1 hour and 5 minutes
Kapalua to Makena: 1 hour and 10 minutes
Kapalua to the Mount Haleakala summit:
        2 hours and 25 minutes
Kapalua to Kahului: 55 minutes
Kapalua to Iao Valley State Park: 55 minutes

Lahaina to Kaanapali: 5 minutes
Lahaina to Kapalua: 15 minutes
Lahaina to Maalaea: 25 minutes
Lahaina to Hana: 3 hours and 20 minutes
Lahaina to Kihei: 35 minutes
Lahaina to Wailea: 50 minutes
Lahaina to Makena: 55 minutes
Lahaina to the Mount Haleakala summit:
        2 hours and 10 minutes
Lahaina to Kahului: 40 minutes
Lahaina to Iao Valley State Park: 40 minutes

PREPARING TO DRIVE TO HALEAKALA NATIONAL PARK

- Start with a full tank of gas. *Gas is **not** available in the park.*
- Leave early!
- Leaving late in the day could mean that the summit is engulfed in cloud cover.
- No food or drinks are available in the park. Bring food, water, juice, or cola with you.
- Bring sunscreen
- Bring warm clothing.
- Day time temperatures can be significantly *cooler* than the temperature at your hotel.
- The weather could be warm, cold, or rainy. Bring clothing for all possible weather.
- Restrooms are available.
- It will take several hours to reach the summit view site.
- *Very approximate* driving times from *west shore resort areas to Kahului:*
    - From Hana: 120 minutes
    - From Kaanapali: 50 minutes
    - From Kihei: 15 minutes
    - From Lahaina: 40 minutes
- These estimates do **not** include traffic delays. Allow more time than these rough estimates!

ROUTE FOR KAHULUI TO HALEAKALA NATIONAL PARK

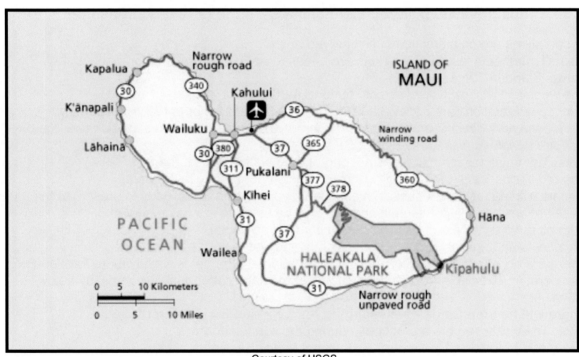

Courtesy of USGS

- To travel to Haleakala's summit, leave Kahului by going eastward on Route 36.

- At the intersection of Route 36 and Route 37, turn right and proceed on Route 37.

- At the intersection of Route 37 and Route 365 , proceed on Route 37.

- At the intersection of Route 37 and Route 377, turn left and proceed on Route 377.

- At the intersection of Route 377 and Route 378, turn left and proceed to Haleakala's summit on Route 378.

TO VIEW THE SUNRISE AT HALEAKALA'S SUMMIT
- All visitors in personal and rental vehicles entering the summit area from **3:00 am to 7:00 am need a reservation.** See the next page for details.
- *Continued on the next page.*

- Parking at the Haleakala's summit is limited. Your reservation ensures you will have a parking space at one of the four summit sunrise viewing locations..
- For a good viewing location, try to arrive at Haleakala's Summit Visitor Center *no later than a half hour before the sun rises. However, your arriving earlier is better.*
- **In the summer, the sun rises as early as 5:38 AM; in the winter, as late as 6:55 AM.**
- Before daybreak, the color of the sky and clouds are stunning.
- Bring warm clothes! It gets very cold at the summit.
- The temperature drops 3 degrees for every 1000 feet of elevation. Thus, at the summit visitor center's 9,740-foot elevation, where most people watch the sunrise, it is about 30 degrees colder than at sea level.
- Be sure to bring pants, shoes, layers of clothing, and blankets. Temperatures are often in the 40-degree range. Don't arrive wearing shorts.
- And, most importantly, don't forget your camera.
- Stay beyond sunrise!
- One of the mistakes visitors commonly make is to leave immediately after the sun rises.
- If you stay ten to twenty minutes later, you will see the dawn colors stretch across the park's landscape.
- Reservations are available **only online**, up to 60 days in advance of your sunrise visit. For reservations, go to:

nps.gov/hale/planyourvisit/haleakala-sunrise-reservations.htm

- Reservations are on a first come, first served basis.
- A small number of last-minute tickets are released **online** two days beforehand at 4:00 PM Hawaii Standard Time.
- The website will show tickets as sold out until 4:00 PM.
- *Sunrise reservations are not available at the entrance station or at the visitor centers.*
- *Calling the park directly, or visiting in-person will not result in a reservation because the park's staff are unable to make reservations for you.*
- Note: The small reservation fee is not part of the park's entrance fee.
- *The regular park entrance fee* will also be collected upon entry to the park.
- Visitors with national park passes, please have your pass and ID ready to present at the gate.
- Reservations are only for sunrise and can only be used on the day reserved.
- Summit conditions cannot be predicted to be clear or cloudy.
- Weather is unpredictable, but it is often windy and wet.
- There are no weather predictions available for sunrise. Sunset is slightly more foreseeable.
- Temperatures immediately before dawn and immediately after dusk are regularly below freezing.
- **There will be no refunds or exchanging reservations for a different day.**
- Note: The visitor centers are not open during the sunrise hours.
- The road up and down the mountain does not have streetlights.
- It takes approximately 1.5 hours to drive from Kahului to the summit.
- Parking is restricted to designated lots only.
- Sunrise parking lots will be closed when full.
- No reservations for sunsets are required at this time (2019).
- Haleakala's "flaming cloud" sunsets are described as breathtaking as its sunrises.

Average Sunrise Times

- The summer and winter average sunrise times are between 5:30 AM and 7:00 AM Hawaii Stand Time.
- Detailed *Hawaii* sunrise times, by day, can be found at:

https://www.bishopmuseum.org/astronomy-resources/

SUMMIT PRECAUTIONS

- At Haleakala's summit you will be at an elevation of 10,020 feet. At this elevation the air is thinner than you are probably accustomed to. Thus:

  - do not walk or run vigorously or you may suffer high altitude induced headaches.
  - the temperature will be cooler at 10,023 feet than at your sea level resort.
  - Bring clothing for all conditions. Be prepared for the summit being warm, cold, or rainy
  - you may need sunscreen. You can get sunburned even on a cloudy overcast day.

- Remember that no gasoline, food, bottled water, or other drinks are sold at the summit, or at the visitor center.
- However, drinking fountains and restrooms are available.

SUNSCREEN

- Beginning in 2021, the State of Hawaii has banded all sunscreen products containing oxybenzone and octinoxate.
- Reason: A 2015 study determined that oxybenzone "leaches the coral reefs of their nutrients and bleaches them white. In addition, it can also disrupt the development of fish and other wildlife".
- These chemicals are used in more than 3500 of the world's popular sunscreens.
- **However,** sunscreen manufactures say that, with in the United States, other ingredients effective for SPF 50 have limited availability.
- **Furthermore,** the Hawaiian Medical Association wants this issue studied more deeply because the evidence suggesting oxybenzone was the cause of coral bleaching was not *peer reviewed.*
- The Hawaiian Medical Association also feared that not wearing any sunscreen would increase cancer rates.
- This issue is left for you to decide.

An Introduction to Haleakala National Park is on the next page

INTRODUCTION TO HALEAKALA NATIONAL PARK

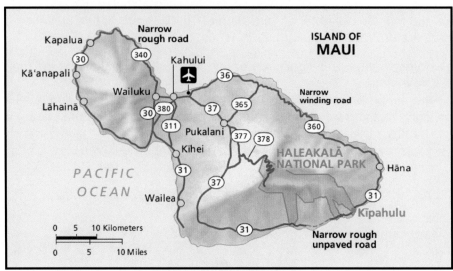

Courtesy of National Park Service

- The location of Haleakala National Park on Maui is shown on the above map.
- This park covers 47 square-miles. *See the map below.*
- Rather than being a single park, it is divided into two sections.
- The Haleakala section goes from sea level up to its 10,023 foot summit area.
- The Kipahulu area extends along the volcano's southeast flank down to the coast.
- However, these two sections are not directly connected by a road.
- The Kipahulu area, at the east end of Maui, is located about 10 miles west of Hana.
- **A detailed map of the park is shown on the next page.**

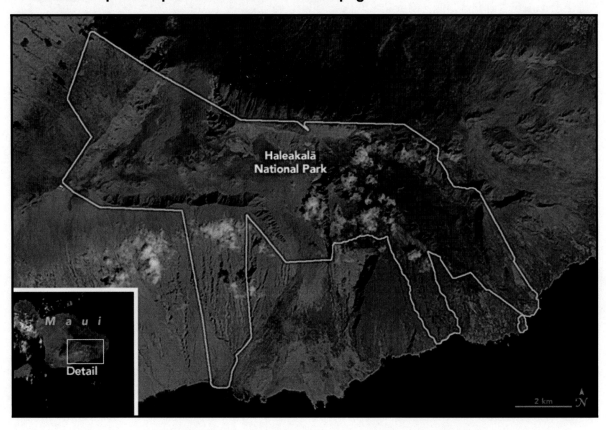

(Courtesy of National Park Service Not to the scale shown)

MAP OF HALEAKALA NATIONAL PARK (Courtesy of the NPS)

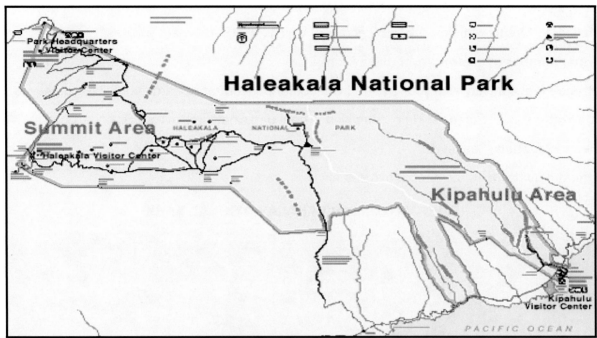

Courtesy of the National Park Service

- For another image of the above map go to **http://npmaps.com/haleakala**

PARK INFORMATION

Mailing Address: Haleakala National Park
PO Box 369
Makawao, HI 96768

Phone: (808) 572-4400 for general park information regarding the Summit Crater Area, Hiking Trails, Camping, Back Country Travel, and Cabin reservations.

A recorded message is available 24-hours a day and will likely contain an answer to your questions.

To speak to a park representative, call the same number and press 0 **during** their office hours of 8:00 am - 3:45 pm Hawaii Standard Time.

For the Kipahulu (Coastal Area) information: Call the Kipahulu Visitor Center at (808) 248-7375 **during** their office hours of 9 am – 5 pm Hawaii Standard Time.

OVER VIEW OF THE PARK

Total area: 47-square-miles or 31,183 acres, of which 24,719 acres are wildness.

Annual number of visitors: over one million.

Entrance fee: can vary with your age and what other park entrance cards you have purchased.

Time zone: Hawaii Standard Time. Hawaii does not participate in Daylight Savings Time.

Lowest elevation: Sea level.

Highest elevation: 10,023 feet above sea level.

RELIEF MAP OF HALEAKALA NATIONAL PARK

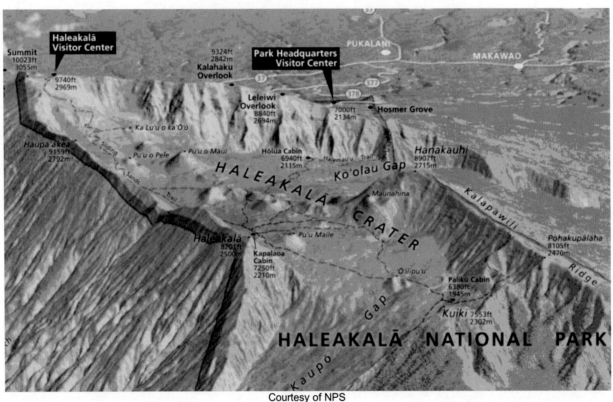

Courtesy of NPS

- As stated on the previous page, *the publisher of this book will not allow maps and diagrams to be shown in landscape view.*
- ***Thus, for a more readable*** view of this image, go to: **http://npmaps.com/haleakala**

ABBREVIATED HISTORY OF HALEAKALA NATIONAL PARK

1916: Hawaii National Park is established by Congress (Big Island Volcano Park and Haleakala combined).

1933-1935: The road to the summit of Haleakala is built.

1934-1941: Early NPS park development by the Civilian Conservation Corps.

1936: The Haleakala Visitor Center was built at the summit.

1937: The backcountry cabins were built.

1941-1946: U.S. Army occupation of Haleakala.

1941 to 1943: The park was closed to the public. • *Continued on the next page.*

1946-present: Additional NPS development in Haleakala National Park.

1958: Park Headquarters was built.

1963: Astronomy observatory was built on Red Hill.

1951: The Kipahula Valley was authorized for inclusion into Haleakala National Park.

1961: *Hawaii National Park* was *separated* and re-designated as *Haleakala National Park* and *Hawaii Volcanoes National Park*.

1962-1978: Nene were re-introduced into Haleakala National Park.

1969: The Kipahulu coastal area of Oheo was included.

1974: The Crater Historic District was listed on the National Register of Historic Places.

1976: Fencing of park boundary began. This fencing was needed to exclude feral animals such as goats and deer in order to protect the park resources. This work continues today.

1999: Kaapahu lands were added to Haleakala National Park.

2008: The Nuu lands were added to Haleakala National Park.

HALEAKALA'S MAJOR SITES

1. **LELEIWI OVERLOOK:** From here, you have a view of the "crater" that is different from the view you will see at the summit.

2. **VIEW FROM THE "CRATER" aka, *PU'U 'ULA'ULA*:** From here, you have a view of looking down into the "crater" with its red and orange cinder cones. Also, some black lava is visible. As described later, this summit is **not crater. Rather it is a basin due to stream erosion.**

3. **SLIDING SANDS TRAIL:** This 5 mile long trail is the park's main hiking trail. It descends over 2400 feet into the "crater" allowing you to have a closer view of the cinder cones and lava flows.

4. **Kalahaku Overlook:** This view site is only accessible as you drive **down from the summit.** This overlook affords another outstanding view of the "crater" floor with its many colorful cinder cones.

5. **Silversword Plants:** This, native to Hawaii plant, is best seen at the summit parking lot where a large number of these plants are located. *See page 18 and 39 for details.*

6. **The Kipahulu section:** A portion of Haleakala National Park located west of Hana. Here you will find lush vegetation, the "Seven Sacred Pools, hiking trails, and waterfalls.

7. **Haleakala's Summit:** *Located up the road from the crater viewing site.* This is where you will find the 10,023 foot elevation sign (which is the highest elevation on Maui). From here you will also have an excellent view of west Maui and the Island of Hawaii.

● **A map of these view sites is on the next page.**

MAP OF THE PARK'S VIEW SITES

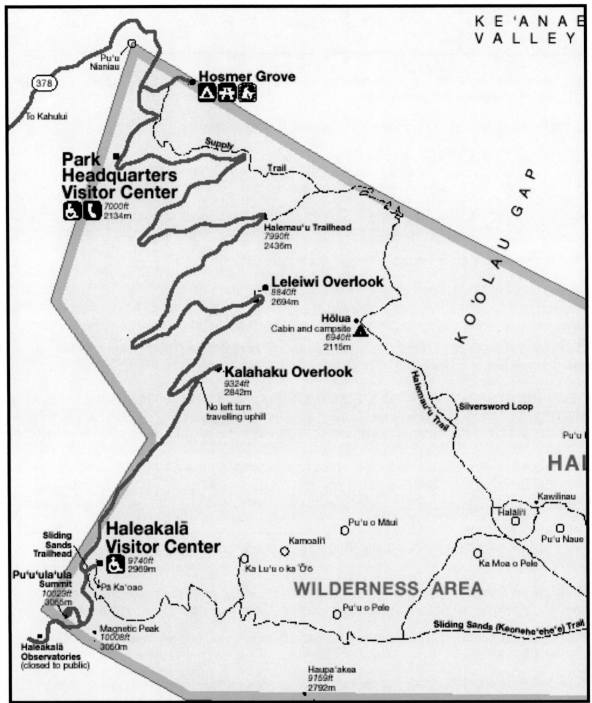

Courtesy of the National Park Service

- **Again**, *the publisher of this book will not allow maps and diagrams to be shown in landscape view*
- Thus, for a better view of this image, go to **http://npmaps.com/haleakala**

THE PARK'S CLIMATE

Temperate data for Haleakala Ranger Station at 6962 ft. (1981–2010)													
Month	Jan	Feb	Mar	Apr	May	Jun	Jul	Aug	Sep	Oct	Nov	Dec	Year
Average high °F	60.6	59.6	59.9	60.4	62.7	65.3	65.6	66.2	64.5	64.0)	63.3	61.3	62.8
Average low °F	43.4	42.3	42.9	43.2	45.1	47.5	48.5	48.8	47.4	47.2	47.0	44.4	45.6
Average rainfall inches	5.99	5.49	7.65	4.17	2.16	1.38	2.59	2.10	2.38	3.10	5.29	7.45	49.75

Source: NOAA (National Ocean and Atmospheric Administration)

HIKING TRAILS (Courtesy of the National Park Service)

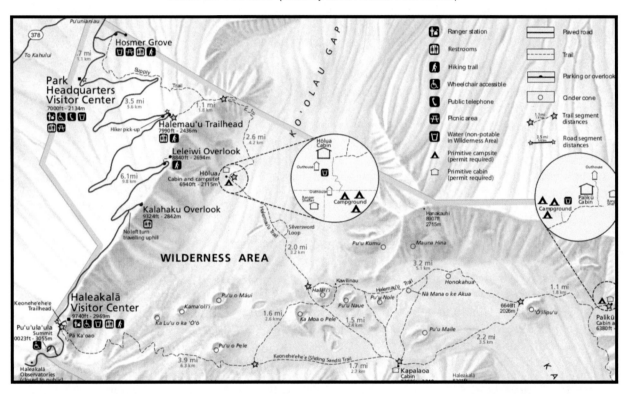

- Again, as stated before, *the publisher of this book will not allow maps and diagrams to be shown in* landscape view. For a readable view of this map, go to **http://npmaps.com/haleakala**

DESCRIPTION OF THE PARK'S HIKING TRAILS

► **THE SLIDING SANDS TRAIL** (also called the *Keonehe'ehe'e Trail):*

- Begins near the Haleakala visitor center and goes into the Haleakala basin. *Photo to the right.*
- Descending 2400 feet over a distance of approximately 5 miles, this hike is the main trail into the "crater".
- This sandy trail is in good condition and offers a closer view of the cinder cones and the lava flows in the "crater".
- *Continued on the next page.*

16

- **Note:** The hike out of the "crater: is steep and hard going. It will take you twice as much time to hike back out than it took you to descend.
- Also, at this high altitude, you should hike slowly, lest you get high altitude sickness.
- Along this trail you will see some silversword plants. *First photo to the right.* These plants are described on page 18 & 39.

- You are urged to talk to the park rangers if you want to hike this entire trail.
- Rental cabins in the basin are available, but reservations are required.
- Sliding Sands Trail's one way distance to:

> **First Overlook**: 0.25 mile with a 50 ft. elevation change.
> **Kapalaoa Cabin:** 5.6 miles with a 2490 ft. elevation change.
> **Paliku**: 9.2 miles with a 3360 ft. elevation change.

▶ THE *PA KAOAO* OR WHITE MOUNTAIN TRAIL:
- This is the other trail that begins near the visitor center.
- This trail starts just outside of the Visitor Center at 9,740 feet.
- White Mountain is a cinder cone volcano. *Second photo to the right.*

- This trail, less than 0.5 miles round trip, leads to the top of *Pa Kaoao* (White Mountain). *Third photo to the right.*
- From the top of this small cinder cone, you will have one of the highest vantage points in the park.
- Stone shelters built by the early Hawaiians that explored Haleakala are still visible from the trail.
- As you walk, look for vesicular and amygdaloidal lava. *See page 53.*

▶ SUPPLY TRAIL FROM HOSMER GROVE TO THE
HALEMAUU TRAIL: A 2.3 mile trail with an elevation change of 975 ft.

▶ HALEMAUU TRAILHEAD (one way) to:
> **Valley Rim:** 1.1 miles long with a 400 ft. elevation change.
> *Holua*: 3.7 miles long with a 1050 ft. elevation change.
> **Silversword Loop:** 4.6 miles long with an 840 ft. elevation change.
> *Kapalaoa*: 7.2 miles long with a 740 ft. elevation change.
> *Paliku*: 10.3 miles long with a 1619 ft elevation change.

▶ LELEIWI OVERLOOK Trail: One way, 0.15 mile with a 50 ft. elevation change.

▶ *PA KA'OAO* Trail: One way, 0.2 mile with a 100 ft. elevation change.

▶ *PALIKU* Trail, one way to:
> **Park Boundary in Kaupo Gap:** 3.7 mi with a 2500 ft. elevation change.
> **Route 31 in Kaupo:** 8.6 miles with a 6130 ft. elevation change.

TIPS FOR HIKING IN HAWAII

1. **Don't keep valuables in your car:** Theft from cars is not uncommon.

2. **Don't hike alone:** You might make a wrong turn and become disoriented.
- *Continued on the next page.*

3. **Be back by nightfall:** The Hawaiian Islands are close to the equator, it gets dark very quickly once the sun sets. Be sure you are not on a trail after the sun sets.

4. **Watch where you swim:** Many island streams and waterfall pools contain a dangerous infectious bacteria called *Leptospirosis,* which can be fatal.

5. **Drink Water:** Prevent dehydration by drinking *before you are thirsty.* Drink fluids before and throughout your hike.

6. **Food:** Keep up your energy. Bring snacks like fruit, a protein bar, or trail mix.

7. **Hiking shoes:** Always wear proper hiking shoes!

8. **Bring a light jacket:** Be prepared for sudden showers and a drop in the temperature.

9. **Bring Sunscreen:** If you have not been in the sun before arriving here, *use sunscreen.* But, even if you already have a tan, y*ou still need to use sunscreen! Skin cancer can be deadly!*

THE NENE (Hawaiian Goose)

- You may see some *nene*, the Hawaiian goose, in the visitor parking lot. *Photo to the right.*
- The *nene* is the official state bird of Hawaii.
- The *nene* (*Branta sandvicensis*) is a species of goose native to the Hawaiian Islands.
- It is found exclusively in the wild on the islands of Oahu, Maui, Kauai, Molokai, and on the Island of Hawaii.
- The Hawaiian name *nene* is derived from its soft call.
- Its species name *sandvicensis* refers to the Sandwich Islands, the original name for the Hawaiian Islands.
- Scientists believe that the *nene* evolved from the Canada goose (*Branta canadensis*).
- This goose probably came to the Hawaiian Islands about 500,000 years ago.
- About 16 inches tall, it is a medium-sized goose that spends most of its time on the ground. But *some sources* say they are capable of flying.
- Females weigh 3.36 to 5.64 pounds, while males average 3.74 to 6.72 pounds.
- Nene's breeding season, from August to April, is longer than that of any other geese.
- Most eggs are laid between November and January.
- Nests are built by females. These eggs are incubated for 29 to 32 days.
- During this time, the male acts as a sentry.
- The nene is a herbivore: eating leaves, seeds, fruit, and flowers of grasses and shrubs.
- The nene, the world's rarest goose, was once common in Hawaii.
- Approximately 25,000 were living in Hawaii when Captain James Cook arrived in 1778.
- However, by 1952, hunting and the introduction of predators: such as the Asian mongoose, pigs, and cats reduced the nene population to 30 birds. But, re-introduction has increased their number. *Please do not feed these birds.*

The Haleakala Ranch
- Haleakala Ranch, primarily situated on the leeward slope of Haleakala, encompasses nearly 30,000 acres from Maui's southern shoreline into Upcountry Maui.
- Incorporated in 1888, five generations of Haleakala Ranch families have strived to preserve the open natural vistas of Upcountry Maui for the benefit of current and future generations.
- As one of the island's large landowners, the ranch is committed to stewardship of Maui's water, land, and other natural resources.
- This ranch's pastures support breeding cows, goats and ewes.
- *Continued on the next page.*

18

- As partners in Maui Cattle Company, they supply the community with healthy, natural grass-fed beef, while practicing holistic grazing management to control invasive species.
- The Ranch's economic diversification includes ecotourism activities such as Skyline Eco-Adventures, Pony Express Tours, and Maui Lavender & Botanicals.
- Currently, Haleakala Ranch is evaluating renewable energy as a new land-use to complement ongoing agricultural operations.
- For information: **haleakalaranch.com/about-haleakala-ranch/haleakala-ranch-history/**

SILVERSWORD PLANTS ON HALEAKALA

- Silversword plants are a rare and endangered species. *See the above photo.*
- In Hawaiian, the silversword plant is called 'āhinahina (literal translation, *"very gray"*).
- This plant, *may have* descended from a single ancestor, the California tarweed, which *may have* reached Hawaii millions of year ago.
- *Most* sources say the Haleakala silverswords (*Argyroxiphium sandwicense* subsp. *macrocephalum*) is part of the daisy family, *Asteraceae*.
- The silversword plant is **not** *restricted just to Haleakala.*
- It is also found from 7,000 to 12,000 feet in elevation on Mauna Loa, Mauna Kea, and on Hualalai on the Big Island of Hawaii
- This plant has evolved to withstand the extreme dryness of the cinder cones on which it grows, and also to the intense sunlight of high elevations.
- Hence, a dense covering of silver hair on its slender leaves helps conserve moisture and protect the plant from the sun's severe rays.
- These plants have one tap root, with shallow, easily damaged smaller roots.
- These smaller roots can extend out as much as 6 feet to collect water.
- The Haleakala silverswords have a short, woody stem 2-3 inches in diameter, crowded with a spiral of thick, dagger-like leaves.
- After growing from 7-20 years, a 3-8 foot high "stalk" with 100-500 flower-heads develops.
- Each flower head has a central disk of hundreds of bright yellow florets, surrounded by a score of short, reddish-purple ray florets. *Photo to the right.*
- After blooming once from June through October, the plant dies.

• A road log for directions to Haleakala's summit begins on the next page

ROAD LOG FROM KAHULUI TO THE JUNCTION OF ROUTE 377 AND ROUTE 378

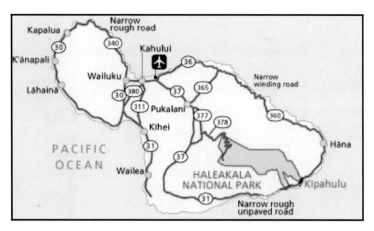

Preview: **Distance traveled: 13.5 miles**

- **First, start with a full tank of gas!**
- *If you proceed to Haleakala National Park, be aware that no gas, food, or water is sold in the park.*
- **Warning:** *Prior to the park boundary, you may encounter bicyclists coming down the road.*
- Almost immediately, Haleakala's *Northwest* Rift Zone cinder cones can be seen on the left side of the highway.
- You will be driving on Kula Volcanics until you get close to Haleakala's summit.
- Refer to the geologic map below and the (Kul) on the geologic map on page 56.
- Until Decembers 2016, 36,000 acres of sugarcane could be seen at the start of this route. But, in 2016, the sugar cane mill went out of business. *See 0.8 miles.*

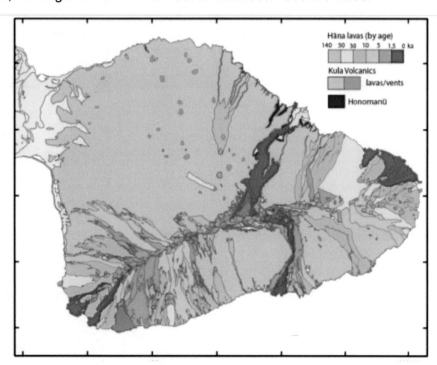

Modified from Sinton 2006

Set your odometer to 0.0 miles at the junction of Route 36 and Route 37.
0.0 miles: Junction of Route 36 and Route 37.
- **Turn right onto Route 37.**
- As you proceed to the intersection of Route 377 and Route 378, the above geologic map and the one on page 56 shows that you are driving over a huge expanse of Kula Volcanics (Kul) erupted 840,000 to 350,000 years ago (Hazlett, 1996 and Sinton 2006).

0.2-1.6 miles: On the left, near the shoreline, a Haleakala *Northwest* Rift Zone cinder cone is visible. *First photo to the right.*

- Looking like "little bumps", several other cinder cones can be seen on the left flank of Haleakala. *Second photo on the right.*

- As the map below shows, Haleakala has three rift zones.

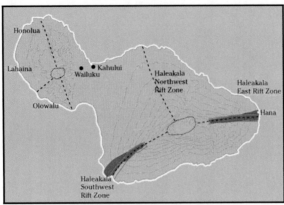

Modified from GeothermEx.Inc

- Rift zones are fractures through which lava is extruded as lava flows, or extrude as pyroclastic debris (ash, cinders, and bombs) to make small cinder cone volcanoes.
- Haleakala's East Rift Zone is its least active rift zone, while the Southwest Rift has had historic eruptions.

0.8 miles: Until December 2016, as far as the eye could see, on both sides of the road, approximately 36,000 acres of a red lateritic soil was under sugar cane production. *Fifth photo to the right.*

- *Note:* At a cost of $30 million system, this sugar cane was irrigated by drip irrigation, making it the largest such system in America.
- But, by December *2016*, after 134 years in business, and sustaining a $30 million loss in 2015, and with the future not looking any better, the Hawaiian Commercial and Sugar Co. closed their mill on Maui (the last sugar mill in Hawaii).

- How this closing would effect the economy of Maui was debated.
- Unfortunately, 675 employees, about half of which had marketable skills, would be terminated.
- However, by *2017*: 326 former workers found new jobs, 25 retired, 19 relocated off island, 48 eight workers were in training programs, and 3 had started new businesses.
- Maui's diversified model calls for the former sugar cane land to be divided into smaller farms with a variety of uses including energy crops, food crops, support for the local cattle industry, and developing an agriculture park.

- But, in 2019, where sugar cane once grew, much of this barren sugar cane land was covered with stubble. *Last photo to the right.*
- But, near the intersection of Route 380 and Route 30, the Pacific Biodiesel Company planted some former sugar cane fields with alternative biofuel crops like sunflowers.

2.4 miles: The massive Haleakala Shield Volcano is visible in the 1:00 direction. Haleakala is called a *shield* volcano because of its shield-like shape. *First photo on the right.*

2.5-3.1 miles: If the sky is clear, looking to the right, *Southwest Rift* zone cinder cones can be seen on Haleakala's right flank. *Second photo to the right.*

3.6 miles: About here, in 2019, some sugar cane is growing because it was still under irrigation.

4.5 miles: Sign: Haleakala National Park: *Third photo to the right.*

6.9 miles: Sign: Historic Makawao Town.
- Makawao is called the biggest "little town" in what is called *Upcountry Maui. See the map to the lower right.*
- Makawao is part plantation town and part art community with small local shops, various boutiques, and art works.
- *Hot Island Glass*, described in websites as the most famous shop in town, is where you can buy glass art work and watch glass blowing.
- However, Makawao is also famous for its Hawaiian cowboys (*paniolos*).
- Since the late 19th century, *paniolos* have tended cattle in Maui's wide-open upland fields.
- And, for more than 50 years, *the Makawao Rodeo,* held on the Fourth of July, is Hawaii's largest paniolos competition.

7.0 miles: Junction of Route 37 and Route 365.
- **Proceed on Route 37.**
- ***WARNING: You will need to turn left at 7.6 miles!***

7.6 miles: Junction of Route 37 and Route 377.
- <u>Turn left</u> **and proceed on Route 377.**

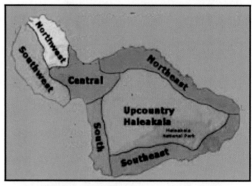

(Map modified from an unknown source).

7.7 miles: Sign: Haleakala Crater 28 miles. *Photo to the right.*
- Actually, Haleakala does **not** have a "crater".
- Haleakala's summit has a basin that is due to *stream erosion. See page 35 and 36 for more details.*
- In this basin, a number of cinder cones are located.

8.6 miles: Elevation 2000 feet (Note: an elevation sign was once here).

8.7 miles: Grove of non-native trees, including *Jacaranda* trees.
- In the summer, these trees can be recognized by their purple blossoms.
- *Jacaranda* refers to a genus of 49 species of flowering plants in the *Bignoniaceae* family that are native to the tropical and subtropical regions of Central and South America, Cuba, Hispaniola, Jamaica, and the Bahamas. However, these trees have also been planted in Nepal, South Africa, Zimbabwe, and Australia.
- Note: As of 2010, 148,530 jacaranda trees were in Los Angles neighborhoods. While their purple blossoms are beautiful, when they fall to the sidewalk below, they make quite a mess.

9.1 miles: Grove of eucalyptus trees planted in 1870 by Ralph Hosmer.
- Ralph Hosmer was Hawaii's first territorial forester.
- He planted groves of pine, spruce, cedar and eucalyptus at different elevations attempting to create a viable timber industry.
- Unfortunately, only 20 of the 86 species he introduced survived because: 1) trees with shallow roots were blown down in storms, or 2) the trees found the soil chemistry unfavorable, or 3) the soil fungi present was unsuitable for the tree's growth or reproduction.
- However, a few trees, such as the Mexican Weeping Pine (*Pinus patula*), Monterey Pine (*Pinus radiata*), and eucalyptus species thrived and migrated from Hosmer's experimental forests.
- Unfortunately, while the eucalyptus trees are pleasingly aromatic, they choke out most of the other native plant-life. Also, their limbs can easily break in strong winds.
- Consequently, these trees have become aggressive invaders. They are now recognized as threats to the native ecosystems within the park.

10.1 miles: Another, large, dense grove of non-native trees.

10.2 miles: Sign: Elevation 2500 feet.
- You started out near sea level. When you reach Haleakala's summit, the elevation will be 10,023 feet; which is twice the elevation of Denver, the *Mile High City.*
- The surrounding pastures indicate that you are in cattle country.
- Also, the road cuts, if not covered with vegetation, display Kula basalts that have weathered to form thick, red lateritic soil.
- In places, you can see the underlying parent basalt.
- *However, proper lighting is necessary to see this soil profile.*

11.4 miles: On the right, *Puu Pane*, a Kula Volcanics cinder cone. *First photo to the right.*
- Like many of the cinder cones on Maui it has been quarried for cinders, which are often used for road construction.

12.2 miles: Sign: Elevation 3000 feet.

12.6 miles: The town of Kula.
- In Hawaiian, Kula means: *plain, field, open country, or pasture.*
- Kula is located in the *Upcountry Region* of Maui. *See the map to the right.*
- Kula is described as a quaint, rustic area on the northwest facing slopes of Haleakala.
- Located at an elevation of 1,600 to 3,600 feet, Kula has the perfect climate to grow almost anything. Hence, this area is the source of most of the island's produce.

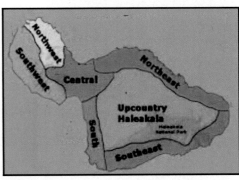

(Map modified from an unknown source).

- Here, without irrigation, Maui farmers grow, herbs, cabbage, lettuce, potatoes, strawberries, coffee, bananas, and tomatoes.
- However, websites state that the sweet Maui onions are cultivated at a lower elevation.
- In addition to vegetables, Kula also cultivates many varieties of flowers such as carnations, proteas, orchids, hibiscus, and jade vines.
- In fact, most of the carnations used in *leis* throughout Hawaii are grown in Kula.
- Maui's upcountry agricultural history dates back to the early Hawaiians who grew taro and sweet potatoes here.
- But, in the early nineteenth century when the whaling fleets arrived, the Hawaiians began to grow Irish potatoes to supply the crews of these ships.
- Today, local farmers continue to adapt to new crops. • *Continued on the next page.*

- Kula is also the center of Maui's culinary resurgence, with much of the exotic produce served at Maui's best regional cuisine restaurants is grown in Kula.
- Kula's most historical landmark is the 1894 octagonal Holy Ghost Catholic Church, a gift from the king and queen of Portugal to the island's Portuguese Catholic plantation workers. .

12.9 miles: The Kula Marketplace is on the right. *Photo to the right.*

- Located in this semi-rural setting, this *gourmet* market is amazing.
- They sell various libations, food, jewelry, ice cream and Hawaiian made items.
- *This establishment is also your last chance to buy food and other refreshments.*
- And, fortunately, *restrooms are located behind the store.*
- But, you need to ask for the code to unlock the restroom doors.

Warning at 13.5 miles, *if you are proceeding to Haleakala's summit, you will need to set your odometer to 0.0 miles.*

13.5 miles: The intersection of Route 377 and Route 378.

Reset your odometer to 0.0 miles at the intersection of Route 377 and Route 378 !!!!!

- **Turn left to continue to Haleakala's summit on Route 378.**
- **Sign: ← Haleakala 22 miles.** *Last photo on the right.*

THE ROAD LOG TO HALEAKALA'S SUMMIT IS ON THE NEXT PAGE.

NOTE: Because other individuals may start *their* trip to Haleakala's summit at a *different starting point,* the material at the beginning of the next road log contains a *brief* discussion of material you read in this road log.

ROAD LOG FROM THE JUNCTION OF ROUTE 377 AND ROUTE 378
TO HALEAKALA'S SUMMIT

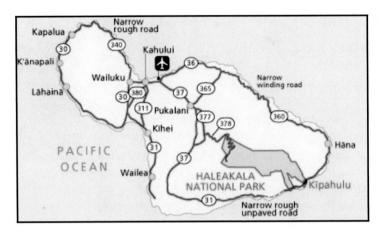

Courtesy of the U.S.G.S

Preview of this route: **Total Distance:** 21 miles.

- As you *may or not have read in a previous road log,* as you climb the western slopes of Haleakala, at an elevation of about 3200 feet, you will be in what is called *Upcountry. See the map to the right.*
- At this elevation, it is a little cooler and east Maui locals feel this is one of Maui's most desirable places to live.
- This area is also the prime agricultural center for many of the vegetables served in Maui's finest restaurants.
- Also, a wide range of flowers for leis and other purposes are grown here.
- Upcountry is also the location of several large cattle ranches and their paniolos (Hawaiian cowboys)
- At Haleakala's summit, you will not see a crater. *Rather you will look into a basin due to* **stream erosion**. Thus, the term "crater" is a misnomer.
- Within this basin, a number of colorful cinder cones can be seen. *Photo to the right.*
- At the summit, hiking trails into the basin will allow you a closer view of these cinder cones.
- Also, a number of astronomical observatories can be seen.
- One of most popular feature of the park is to view the sunrise from the summit.
- **But Reservations are required!** *See page 7 for critical information!*
- Also, in the park, you will see silversword plants and possibly a nene (The Hawaiian goose).
- The geologic map on the next page shows that as you drive from Kahului to the park, you will drive *mainly* over Kula lava flows, extruded 950,000 to 150,000 million years ago (USGS).
- The Kula lavas flows are thin and parallel the slopes. They are often interbedded with cinders and ash, or buried just in ash.
- As you drive upward, road cuts will show this sequence of interbedded cinders and lava a number of times.
- The dominate Kula lava is hawaiite, a basalt (i.e., one rich in sodium) that contains green olivine grains.
- Note: Hawaii jewelry stores sell the gem variety of olivine as the birthstone peridot.

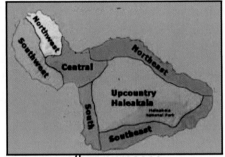
other purposes.

Courtesy of alohaisles.com

- *Continued on the next page.*

GENERALIZED GEOLOGIC MAP OF MAUI

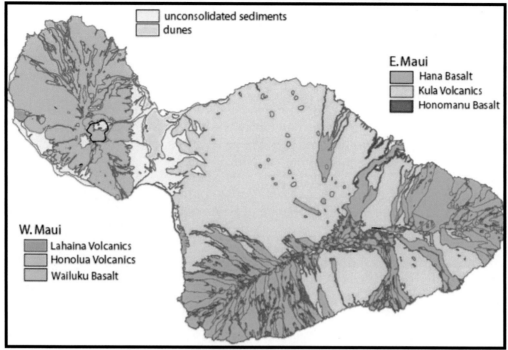

Modified from Sinton 2006

BEGIN THE ROAD LOG TO HALEAKALA NATIONAL PARK FROM THE JUNCTION OF ROUTE377 AND ROUTE 378

Set your odometer to 0.0 at the intersection of Route 377 and Route 378. *Note: odometers and your starting point can vary by 0.1-0.2 miles.*

0.0 miles: The junction of Route 377 and Route 338.
- **Turn left onto Route 378.**
- **SIGN:** Haleakala National Park 22 miles. *First and Second photo to the right.*

0.3 miles: On the right **Sign: elevation 3500 feet.** *Third photo to the right.*
- You are beginning your climb up to an elevation of 10,023 feet!
- In these 22 miles, there are 32 switchbacks as you drive up a road with an average 6% grade (Decker and Decker, 1992)
- ***Beware of bicycle riders coming down this road!***

1.7 miles: Sign: elevation 4000 feet. *Last photo to the right.*

2.3 miles: On the left, a zip line attraction, and on the right a lavender farm.

2.5 miles: Grove of eucalyptus trees.
- *As you may have read earlier,* these eucalyptus tree were first planted on Maui in 1870 by Ralph Hosmer.
- Hosmer was Hawaii's first territorial forester who planted groves of pine, spruce, cedar, and eucalyptus at different elevations attempting to create a viable timber industry.
- *Continued on the next page.*

- Only 20 of the 86 species he introduced survived because 1) trees with shallow roots were blown down in storms, and 2) other trees found the soil chemistry or the fungi present unsuitable for growth or reproduction.
- However, a few trees, such as the Mexican Weeping Pine (*Pinus patula*), Monterey Pine (*Pinus radiata*), and eucalyptus species thrived and migrated from Hosmer's experimental forests.
- Unfortunately, while the eucalyptus trees are pleasingly aromatic, they choke out most of the other native plant-life. Also, their limbs can easily break in strong winds.
- Consequently, these trees have become aggressive invaders, and they are now recognized as a threat to the native ecosystems within the park.

2.7 miles: Sign: Open Rangeland. *Photo to the upper right.*
- Looking around you can see the wide expanse of Haleakala Ranch cattle pasture lands

3.8 Don't stop here, stop at 4.0 miles.

4.0 miles: From this view site, if the sky is clear, you can see the West Maui Volcano (*photo to the far left below),* Maui's Isthmus (center *photo below*), Kahului on Maui's east coast (*photo to the far right).*

Also visible is Kihei on Maui's west coast (*far left photo below*),
Kahoolawe *(center photo below),* Lanai (*far right photo below),*

- Also visible is the Molokai Islet *(photo to the far left),* astronomical observatories on Haleakala's summit (*center photo*), and looking back to the right, cinder cones on Haleakala's Southwest Rift Zone (*photo to the far right*).

- *Continued on the next page.*

THE MAUI ISTHMUS

- The Maui Isthmus is located between Haleakala and the West Maui Volcano. It is bounded on the west by Kihei and on the east by Kahului
- This flat, fertile area is why Maui is called the *Valley Isle.*
- This isthmus formed during Haleakala's shield building stage as its lava abutted into the West Maui Volcano.
- Also, stream sediments from the West Maui Volcano and Haleakala were also deposited here.
- These sediments and lava flows were chemically weathered to form the soil on which sugarcane was grown until December 2016.
- Haleakala has three rift zones. *See the map to the right.*
- Looking across the road, a red lateritic soil is in the road cut
- Lateritic soils form in hot, wet tropical climates where intense chemical weathering can decompose basalt or sediments into soil.
- The red color of this soil is due to the iron bearing minerals in the parent material chemically altered to iron oxide.
- This iron oxide then stains the surrounding soil particles red.

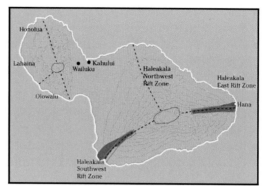

Modified from GeothermEx.Inc.

4.8, 6.2, and 6.8 miles: All of the road cuts at these mileage points show the same thin Kula lava flows interbedded with pyroclastic debris (ranging in size from coarse to fine-grained material). *First, second, and third photos to the right.*
- Pyroclastic fragments are produced by violent eruptions that throw blocks, cinders, ash, and bombs into the air.
- *Pyro* comes from the Greek meaning fiery, and *klastos* from the Greek meaning broken.
- Some of the Kula lavas are vesicular (i.e., they contain gas bubble holes).
- These gases are dissolve in the subsurface molten material called *magma.*
- When the magma is extruded as *lava,* the confining pressure is reduced; and the dissolved gases escape forming vesicles i.e., gas bubble holes. *Fourth photo to the right.*
- These rocks are then called vesicular basalts.
- Also, these thin lava flows are indicative of lavas extruded onto steep slopes

5.0 miles: More grass covered Kula lava.
- This grassy vegetation allowed large cattle ranches to develop in this area.

6.2 miles: Kula blocky lava flows interbedded with ash from *Puu Pahu,* a Kula cinder cone (Sinton, 2006).
- *The Puu Pahu* cinder cone will be seen at 7.2 miles.

6.6 miles: On the right, a view site with the same features seen at 4.0 miles. But, from here, if the sky is clear, you have good views of the West Maui Volcano and Maui's western or leeward shoreline.

28

6.7 miles: Same lava and cinders as seen earlier.
- The road keeps cutting through the same rock sequence of lava interbedded with pyroclastics. *First photo to the right.*

7.1 miles: Road cuts in red ash from *Puu Pahu (see 7.2 miles).*

7.2 miles: On the left, a scenic turnout.
- Pull-in and look across the road and upslope.
- The *Puu Pahu* cinder cone is visible. *Second photo to right.*
- Also, in the road cut across the road, red ash from the *Puu Pahu* cinder cone is present.
- *Pukiawe (Styphelia tameiameiae)* bushes are growing abundantly on the slopes of this cinder cone. *Third photo to the right.*
- This plant ranges from sea level to the alpine zone.
- But, it will grow in decreasing density as you approach Haleakala's summit.
- Its berries are an important food source for the nene, the Hawaiian goose. *Fourth photo to the right.*

8.2 miles: *SIGN:* Saving Haleakala's Sub-alpine shrub land.

9.1 miles: Another grove of eucalyptus and other evergreens

9.2 miles: SIGN: elevation 6800 feet.
- On the left, you can see *part* of the *Puu Nianiau* cinder cone. *Puu* meaning hill or peak, and *Nianiau* meaning *swordfern.* .
- This cinder cone was once thought to be a *Kula Volcanics* cinder cone, but a newer K-Ar age of 113,000 years indicates it is an early Hana cone (Sinton, 2006).
- Because of quarrying, the dipping layers of red and black ash can be seen in cross-section. *Fifth photo to the right.*
- Layers of volcanic cinders and ash ejected out of cinder cone were deposited on its slopes as the cinder cone enlarged. *Sixth photo to the right.*
- **Cinder cones:**
  - ▸ are the most common type of volcanoes.
  - ▸ are symmetrical, cone shaped volcanoes with a bowl-shaped summit. *Diagram to the lower right.*
  - ▸ may occur as a single volcano or as secondary volcanoes on the sides of shield volcanoes.
  - ▸ form around a single volcanic vent where lava explodes into the air and cools very quickly to form a ring of accumulating cinders, ash, and bombs (aka, pyroclastic debris).
  - ▸ are not very tall because nothing is cementing the pyroclastic fragments together.
  - ▸ are steep sided, with slopes never exceeding the angle of repose, 32-34 degrees (the maximum stable angle to which dry loose material can accumulate).
  - ▸ Basaltic lava flows may breach cinder cones, or the lava may escape by flowing under the cone.

  Note: Cinder cone diagram to the right is modified from Explore Volcanoes .com.

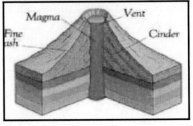

10.0 miles: In the 11:00 direction, an astronomical observatory may be visible.

10.1 miles: You have reached the boundary of Haleakala National Park. *First photo to the right.*

- **And,** the Haleakala Park pay station. *Second photo to the right.*
- Also, you are entering the **Northwest Rift Zone**, one Haleakala's three rift zones. *See the map below.*

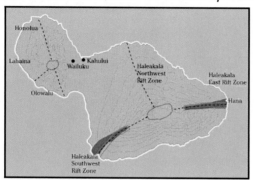

Modified from GeothermEx.Inc.

- The other two rifts zones are the **East Rift Zone** and **Southeast Rift Zones**.
- The **East Rift Zone** is the least active of the three rift zones.
- **Northwest Rift Zone** cinder cones can be seen from the Route 36/Route 37 interchange.
- **Southwest Rift Zone** cinder cones are visible from the Maui Isthmus, or from the west coast of Maui.
- **Southwest Rift Zone** cinder cones produced Haleakala's most recent eruptions (from 1480 to1600).
- At **11.0 miles,** you will reach the **Haleakala Park Headquarters and Visitor Center.**
- Then, *9.3 miles from the visitor's center*, you will reach the rim of Haleakala's summit *basin* (aka, crater) and another visitors center.
- As you drive to the summit, note the dramatic decrease in vegetation as the altitude increases

10.2 miles: On the left, the road to the Hosmer Grove Campground. *First photo to the right.*

- This camp grounds, just below 7000 feet, has daytime temperatures *averaging* 50 to 65° F (10-18° C), with the night temperatures dropping to near freezing (0° C).
- Picnic tables, BBQ grills, drinking water, and pit toilets are available.
- A self-guided nature trail begins and ends at the campgrounds.
- This camp ground was named after Ralph Hosmer, who was previously discussed at 2.5 miles.
- An astronomical observatory is visible in the 12:00 direction.

11.0 miles: SIGN: Elevation 7000 feet. *Photo to the right.*

- Here, a very dense vegetative cover is present.
- In addition to rain, the seemly always present clouds bring additional moisture to these plants.

11.1 miles: Haleakala Park Headquarters and Visitor Center.
First and second photos to the right.

- At 7000 feet above sea level, you are at the park's headquarters and visitor center.
- As the elevation increases, you need to watch your body's reactions to the thinner air at these higher elevations.
- Do not try running or exerting yourself until you are acclimated to these elevations.
- When you reach Haleakala's summit, you will be at an elevation of 10,023 feet.
- At this elevation, you are almost 2 miles above sea level.
- Being too active can cause severe headaches.
- Restrooms are available to the left of the center's front door.

- Also, follow the traveler's creed: *Never stand when you can sit, and never sit when you can lie, and **never** pass up a restroom stop!*
- In February, the average high temperature at the visitor center is 59°, and the average low is 47°F.
- In August the average high is 66°, and the average low is 47°F. (Doughty 2012).
- In contrast, at the summit, it's about ten degrees cooler, which at times, receives a light snowfall.

WHEN YOU READY TO LEAVE. RESET YOUR ODOMETER TO 0.0 MILES AT THE INTERSECTION OF THE PARKING LOT'S EXIT ROAD AND ROUTE 378.

THE ROAD LOG THE HALEAKALA'S SUMMIT CONTINUES ON THE NEXT PAGE.

0.0 miles: INTERSECTION OF THE VISITOR CENTER PARKING LOT EXIT ROAD AND ROUTE 378.

- Turn right and proceed to Haleakala's summit.
- Another cinder cone is directly ahead. *First photo to the right.*

0.6 miles: Abundant grassy and brushy vegetation is growing on soil that has formed from the weathering of Kula lava or ash.
- Again, the clouds are supplying additional moisture.

1.2 miles: Road cut in massive Kula lava.

1.4 miles: On the left, a red laterite soil is between two lava flows. *Second photo to the right.*
- A considerable period of time was needed for the lower lava flow to be chemically weathered to produce this soil.
- Then, later, this soil was buried by the overlying lava flow.

1.7 miles: On the right, and in the distance, the *Puu Oili* cinder cone is visible with a plethora of antennas. *Third photo to the right.*
- This same cinder cone will be seen several more times.

1.9 miles: Road cut in a vesicular Kula basalt. *Fourth photo to the right.*
- *But, it is necessary to exit your car to see the vesicles.*
- As stated earlier, gases are dissolved in a magma (i.e. the subsurface molten material).
- Then, when the lava is erupted on the surface, the pressure is reduced, and the dissolved gases escape making holes as the lava solidified.
- These gas bubble holes are called vesicles, and the resulting rock is called a vesicular basalt.
- Also, a very dense growth of vegetation covers the slopes.
- But as the elevation increases, all vegetation will become less and less abundant.

2.0 miles: Note the abundance of the *pukiawe* bushes.
- These plants will decrease in abundance as you go towards the summit.

2.4 miles: More massive Kula lavas.

3.3 miles: On the right, again you can see lava overlying a red laterite. *Fifth photo to the right.*
- Again, laterites are soils that form by intense chemical weathering from sediments or basalts in tropical climates.

3.4 miles: Sign: 8000 feet in elevation.
- **Halemauu Trail trailhead,** on the left. *Last photo to the right.*
- This trail goes into the Haleakala basin.
- Ahead, you will see many road cuts of lava inter-bedded with cinders, and lava resting on a red lateritic soil.

4.5 miles: Fewer *pukiawe* bushes are present.

5.6-5.7 miles: On the right, the ground is covered with brown ash supporting the growth of *pukiawe,* grass, and other plants that, as previously stated, are becoming less abundant. *First photo to the right.*

- Unlike passive eruptions that produce basaltic lava, violent cinder cone eruptions produce cinders and ash.

6.0 miles: Sign: Leleiwi Overlook. *Second photo to the right.*
- At the end of this trail, you can see into Haleakala's basin.
- **The parking lot is on the right. It comes up quickly!**

6.1 miles: On the right, the **Leleiwi Overlook Parking Lot**.
- Cross the road and walk the 0.15 mile Leleiwi Overlook Trail.
- **Warning:** in places this trail is *rough and irregular.*
- Along this trail, many of the lavas are vesicular.
- A late 1880's stone wall, once 2 miles long, is at the start of this trail. *Third photo to the right.*

- Ranchers built this wall to move cattle from Haleakala's rugged terrain to the east and west sides of Maui, where grassy pastures are found.
- Unfortunately, the cattle had a deleterious effect on the natural vegetation.
- Consequently, the park constructed 30 miles of fencing to protect the park's unique plant ecosystem. However, deer, goats, and pigs also threaten the parks vegetation.

VIEW FROM THE LELEIWI VIEW SITE

- The 3000 foot deep, 7.5 mile long, and 2.5 mile wide Haleakala basin (Sinton 2006) can be seen from the end of this trail. Numerous, colorful cinder cones are located in this basin..
- Decker and Decker (1999) state that Manhattan Island could fit inside this basin.
- *Haleakala does **not** have a crater at its summit.*
- Rather, this basin formed by the Koolau and Kaupo valleys eroding back their valley walls in a process called headward erosion.
- Eventually, this headland erosion breached Haleakala's summit.
- The *drawing below* shows Haleakala's **stream eroded basin** with the stream eroded gaps.
- In this sketch, the **Koolau Gap** is at the top of the basin.
- The **Kaupo Gap** is at the bottom of this sketch,.
- Sketches of numerous cinder cones are also shown in this basin.

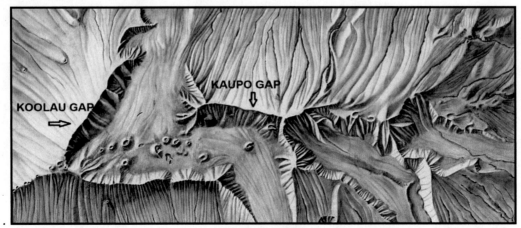

Modified from NPS

- *Continued on the next page*

- *Unfortunately*, sometimes in the late afternoon, clouds may block your view of these cinder cones.
- Also, in the mist that can form around your shadow, a halo of colors may form.
- This halo effect, called a *Specter of Brocken*, is due to small spaces between the cloud's water droplets diffracting the sunlight.

OPTIONAL GEOLOGIC INFORMATION AT THE LELEIWI SITE

- Returning to the parking lot, face the road cut and walk about 100 feet *down* the road to a road cut with a complex lava flow. *Lower left photo.*
- The upper several feet of this flow, which overlies ash, contains abundant *pea size* green olivine grains and *large black crystals* of pyroxene. *Lower right photo below.*

- According to Sinton (2006), these minerals indicate that this flow is a rare variety of alkalic basalt called ankaramite (that is, the lava contains above average amounts of sodium or potassium).
- However, the lower portion of this outcrop contains *few large* pyroxene crystals.
- A theory to explain a lava flow with large crystal overlying a fine-grained lava flow states that the initial eruption tapped the *upper part* of the magma chamber containing the small crystals.
- Then, the lower part of the magma chamber erupted where the large crystals had settled.
- As a result, the large crystals in the lower part of the magma chamber are *overlying the upper part* of the magma chamber where the smaller crystals are located.
- For more details go to:
  www.soest.hawaii.edu/GG/resources/docs/Maui_2006.pdf

6.8 miles: Sign: Elevation 9000 feet. *Photo to the upper right.*
- A big change in the vegetation has taken place.
- This landscape is one of boulders with scattered *pukiawe* plants growing between them. *Photo to the right.*

7.2 miles: The view from here is similar to that seen in a previous road log at 4.0 miles.
- On the right, *if the sky is clear*, you can see the West Maui shoreline and the islands of Lanai, Molokai, Molokini, and Kahoolawe. *See the map to the right..*
- Molokini, not shown on this map, is an eroded volcano islet located in the channel between Maui and Kahoolawe.

Courtesy of the United States Geological Survey

34

7.6 miles: THE KALAHAKU OVERLOOK.
- **NO LEFT TURN ALLOWED!**
- *Go to this view site on the <u>way down</u> from the summit area.*
- *Then, as you descend from the summit, you can turn right.*
- The Kalahaku trail is a ***must*** see site!
- *This view site is described at the end of this road log.*
- Haleakala's summit is 2 miles ahead.

7.9 miles: On the left, lava is "draped" over older tan colored cinders. *First photo to the right.*
- Ash will become more abundant as you drive upward.

8.9 miles: Several distant summit astronomical observatories are visible in the 1:00-2:00 direction. *Second photo to the right.*
- Also, the vegetation is sparse, with just a few plants growing between lava boulders on an ash soil. *Third photo to the right.*

9.0 miles: From here, on the *far left*, you can see the White Mountain cinder cone. *Photo to the lower left.*
- *Turning to the right*, you can see a large cinder cone. And, behind this cinder cone, in the far distance to the left, Red Hill, the highest elevation on Maui (10,023 feet) is visible. *Middle photo below.*
- *Turning farther right*, a cinder cone with astronomical observatories is visible. *Photo to the lower right.*
- Finally, on the far right antennas can be seen on another cinder cone.

9.4 miles: Entrance to the Haleakala Visitor Center parking lot and view of the summit basin, aka the "crater."
- On the left, entrance to the Haleakala Park Visitor Center. *Photo to the right.*
- Elevation: 9740 feet. But, the highest elevation in the park is at the highest summit parking lot.
- Haleakala means: *"House of the Sun".*
- This name comes from the demi-god Maui, and how he captured the sun.
- Please refer to page 41 for the legend of Maui and his capturing the sun.
- Ancient Hawaiian structures and a heiau (altar) have been found on Haleakala's rim. Thus, early Hawaiians must have thought this area was a very special and important place.
- Also, a fine-grained Kula basalts, found in a nearby quarry, was favored by the ancient Hawaiians for making adzes (an axe-like tool with a curved blade positioned at right angles to the handle, used for shaping wood).
- *Continued on the next page.*

- Ancient Hawaiians also came here to bury the umbilical cord of a child to protect it from rats (Doughty, 2012).
- As stated earlier, tourists come here to view the sunrise. *See page 7.*
- The visitor center located here has geological, geographic, and biologic displays; *but it is only open from sunrise to noon.*
- But, the adjacent restrooms are open 24 hours.

FROM HALEAKALA'S VIEW SITE

- From here you can look down into the nearly 7.5 mile long and nearly 2.5 miles wide Haleakala Basin (Sinton, 2006). *Photo to the right.*
- *As stated earlier, people call this 3000 foot depression a crater, but it is not volcanic in origin.*
- *Rather, it is a stream erosional basin due to streams eroding back into their headwalls.*
- Stream erosion probably formed this basin about 145,000-120.000 years ago.
- This basin contains a number of cinder cones *Second photo on the right.*
- The crude line of colorful Hana cinder cones are a continuation of Haleakala's Southwest Rift cinder cones that crossed the summit into the Haleakala's East Rift Zone. *See the diagram below.*

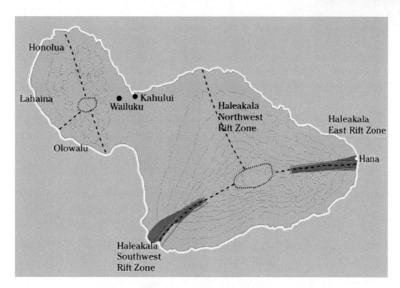

Modified from GeothernEx.Inc.

- The youngest volcanic features in Haleakala's basin are the 4500 year old Hana Volcanics (Sinton, 2006).
- These Hana cinder cones have bright red and orange colors due to iron bearing minerals being oxidized by steam and other volcanic gases extrude during and after the cone's eruptions.
- The illustration on the **next page**, *derived from a National Park Service Haleakala display*, identifies many of the features seen in the Haleakala basin. Thank you NPS!

- *Continued on the next page.*

....

36

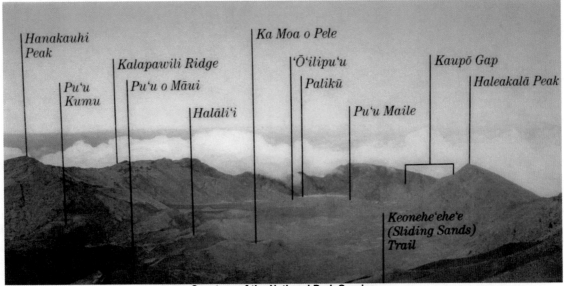

Courtesy of the National Park Service

- From the basin floor, the tallest cinder cones are 400-600 feet tall.
- Haleakala Peak, seen *on the far right in the above illustration*, has an elevation of 8278 feet.
- The green rocks of the basin's wall are composed of Kula Lava. *Photo to the lower left.*

- *Illustrated in the drawing below,* the rim of this basin is cut by two gaps.
  ▸ **The Koolau Gap** is shown on the northern or upper side of the basin. *Also see the above middle photo.*
  ▸ **The Koolau Gap**, at the head of the Keanae Valley, and on Route 360 (the road to Hana).
  ▸ **The Kaupo Gap** is shown on the south or lower side of the basin. *See the upper right photo above.*
  ▸ **The Kaupo Gap** is also visible in the Kaupo area on the south side of the island.

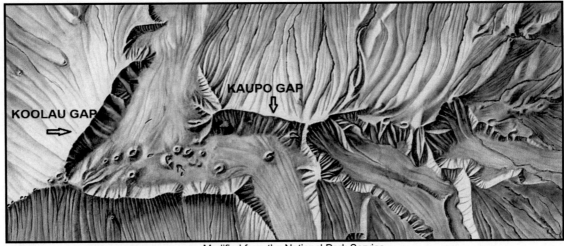

Modified from the National Park Service

- *Continued on the next page.*

ADDITIONAL FEATURES SEEN FROM THE SUMMIT VIEW SITE

The Pa Kaoao Trail:

- Behind you is the **White Mountain** cinder cone and the *Pa Kaoao* Trail. *First and second photo to the right.*
- The *Pa Kaoao Trail,* less than a 0.5 mile round trip hike, leads to the top of *Pa Kaoao* (aka, White Mountain), one of the many small cinder cone volcanos on the slopes of Haleakala.
- This trail offers one of the highest vantage points in the park and gives views of the volcano's wilderness landscape.

- Stone shelters, built long ago by the early Hawaiians that explored Haleakala may be visible from the trail.
- Also, websites state that standing at the top of White Mountain, you *might be at one of the park's less crowded,* sunrise viewing sites.

The *Keoneheehee* (Sliding Sands) Trail (second photo to the right)**:**

- Guide books say that the **Keoneheehee** (Sliding Sands) Trail is the best Haleakala summit hike.
- However, due to the summit's elevation and the steepness of the terrain, this is a very strenuous trail (descending 2,800 feet in the first 4 miles to the valley floor).
- This popular 11 mile trail begins near the entrance to the Haleakala Visitor Center parking lot (elevation *about* 9760 Feet).
- This hike, which takes a full day, crosses the valley floor and ends at the Halemauu Trail at 7,990 ft.
- It is recommended that you have a car meet you up at this point or that you have a car parked in place.

Points of Interest seen on the Sliding Sands Trail

- The *Ka Luu o ka Oo* **cinder cone** is 2.5 miles down the trail.

- The **"crater floor"** is 3.9 miles down this trail, your change in elevation of almost 2,500 ft.

- **"Pele's Paint Pot"** is about 5.7 miles down the trail. This location, near the north side of the *Halalii* cinder cone, is roughly the halfway point of this hike.

- *Kawilinau* **Cinder Cone**, also about 5.7 miles down the trail was formerly called the *"bottomless pit.* But actually it is only 65 feet deep.

Want a Shorter Less Strenuous Hike?

- For a less strenuous half-day hike, walk the first 2.5 miles of the Sliding Sands Trail down to the *Ka Luu o ka Oo* cinder cone. The gain or loss between the trail head and the cinder cone is approximately 1,400 ft
- Once at the *Ka Luu o ka Oo* cinder cone, turn around and hike back out for a total of 5 miles.
- However, it might feel longer due to the extreme elevation, and the fact that you are hiking back up hill.

Other hiking trail information can be found on page 15 and at the visitor center.

FOR A TOPOGRAPHIC MAP OF THE PARK AND THE SUMMIT BASIN GO TO:

Peakbagger.com Large Map Page - Haleakala Crater-South Rim

Then scroll down to the lower right to view this map.

BEYOND THE VISITOR CENTER

When you are finished at the summit basin, exit the parking lot, set your odometer to 0.0 miles at the intersection of the parking lot exit road and the park highway.

0.0 miles: Intersection of the summit basin parking lot exit road and the park highway.
• Turn left.

0.1 miles: Now, all vegetation ceases. All that is present is ash.
• **Also, Magnetic Peak** is on the left. *First photo to the right.*
• Magnetic Peak, Maui's second highest point with an elevation of 10,008 feet, is a geological oddity.
• Because this is an iron-rich cinder cone, it has a magnetic field strong enough to deflect a compass needle.
• And, as seen elsewhere, oxidation of the iron rich cinders gives Magnetic Peak a rusty color.

0.6 miles: Enter the summit parking lot.
• Once at the summit parking lot, you are standing on the *Pu'u 'Ulaula* cinder cone.
• Walking up to the summit observation station, you will be at the park's highest elevation of 10,023 feet. *Bottom three photos below.*

• Your elevation is twice the elevation of Denver, the *mile high city.*
• A number of informative signs are located at this observation view site.
• From this location, *if the sky is clear,* you can see the Island of Hawaii. *Photo to the right.*
• Also, signs along the trail that circles the parking lot provide information regarding the astronomical and Air Force facilities located here. *See Science City below.*
• Also, a large planting of silverswords plants are in this parking lot. *Photo to the right.*
• ***In this photo***, you can see this plant in three stages: living on the right, blooming in the left, and dead in the lower left corner. *Details regarding this plant are given on the next page.*
• On the hill top opposite the observation station, *if the sky is clear*, you can see the West Maui Volcano.

SCIENCE CITY

• Officially, Science City is know as the *Haleakala High Altitude Observatory. Note: this facility is not open to the public.*
• Science City is located on Haleakala's summit because of the elevation, the absence of lights from a major city, and the dryness and stillness of the air. Thus, this site is a prime astronomical location. *Last photo to the right.*

• *Continued on the next page.*

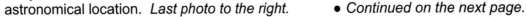

- This facility is operated by the University of Hawaii, the Smithsonian Institution, the United States Air Force, NASA, and others.
- The instruments located here include optical, radio, radar, and infrared telescopes (as well as listening dishes and sensing instruments).
- On a 24/7 basis, these various observatories explore the universe, monitor natural and manmade space activities, and track the growing amount of space junk.
- This entire facility is supported by labs and computing facilities far below at the *Advance Technology Center in Pukalani and at Kihei's Maui High performance Computing Center (MHPCC).*
- Also, instruments at this facility confirmed the theory of plate tectonics when measurements showed that Maui moves 4 inches to the northwest each year!
- For a complete description of the observatories located here, go to:

http://www.ifa.hawaii.edu/haleakalanew/observatories.shtml

SILVERSWORDS PLANTS

- Silversword plants, a rare and endangered species, can be seen in the enclosure in this parking lot.
- In Hawaiian, the silversword is called *'āhinahina* (literal translation, *"very gray"*).
- This plant, *may have* descended from a single ancestor, the California tarweed, which *may have* reached Hawaii millions of years ago
- *Most* sources say the Haleakala silverswords (*Argyroxiphium sandwicense* subsp. *macrocephalum*), is part of the daisy family, *Asteraceae.*
- This plant is **not** restricted just to Haleakala. On the Island of Hawaii, it is also found from 7,000 to 12,000 feet on Mauna Loa, Mauna Kea, and Hualalai,
- This plant has evolved to withstand the extreme dryness of the cinder cones on which it grows, and also the intense sunlight at high elevations.
- A dense covering of silver hair on its slender leaves helps conserve moisture and protect the plant from the sun's severe rays.
- These plants have one tap root and shallow, easily damaged smaller roots that can extend out as much as 6 feet to collect water.
- The Haleakala silverswords have a short, woody stem, 2-3 inches in diameter, crowded with thick, dagger-like leaves arranged in a spiral.
- After growing from 7-20 years, a 3-8 feet high "stalk" with 100-500 flower-heads develops.
- Each flower head has a central disk of hundreds of bright yellow florets, surrounded by a score of short reddish-purple ray florets.
- After blooming once, for about 2 months from June through October, the plant dies.

AFTER YOU ARE FINISHED AT THE SUMMIT AREA

- **Return to your car and drive 2.2 miles down hill to the Kalahaku Overlook.**
- Turn right and drive 0.3 miles to the overlook's parking lot.
- Then walk to the end of this trail for a different view of the cinder cones and lava in the summit basin.
- Informational signs give you're the names of the cinder cones in the basin, along with other details.   • *Continued on the next page.*

Kalahaku Overlook

- Looking down from the Kalahaku Overlook, the basin floor is 2000-2500 feet below.
- *To your left, the Koolau Gap can be seen. See the above photo.*
- *As described earlier*, the basin's rim is cut by two gaps: the *Kaupo Gap* on the rim's far southern side, and the *Koolau Gap* on the on the rim's northern side. *See the illustration below.*

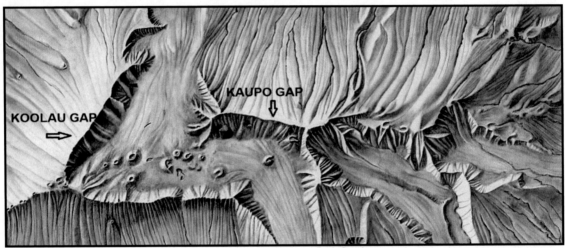

Modified from the National Park Service

- The *Kaupo Gap* was eroded by the *Kaupo Stream*, while the *Koolau* was eroded by the *Keanae Stream*.
- After these two streams eroded deep canyons and valleys, a considerable amount of time passed.
- Then, a period of renewed volcanic activity occurred filling the valley floors with lava and ash to create their present valley floors.
- *At one time,* this volcanic activity was described as a rejuvenation stage in Haleakala's development.
- *However, new dating has shown that these eruptions are Hana volcanics, and that Haleakala is near its post-shield growth stage and **not** in a rejuvenation stage (Sherrod, et al, 2003).*
- Also, *if the sky is clear*, looking across this basin, in the far distance, you can see the Big Island of Hawaii and the summits of Mauna Loa and Mauna Kea.

....

HALEAKALA AT ITS PRIME

- Today, Haleakala has an elevation of 10,023 feet, which is also the highest elevation on Maui.
- But, was Haleakala ever taller?
- All of the Hawaiian volcanoes have been reduced in elevation by erosion and by subsidence, that is, the sinking of the earth's crust.
- The earth's crust cannot support heavy loads of ice, sediment, or rock. Thus, under the massive weight of the huge Hawaiian volcanoes, these volcanoes have subsided.
- Consequently, Haleakala achieved its maximum height about one million years ago at the end of its shield building stage.
- The top of Haleakala's shield, at an elevation of about 7000 feet, is now buried by younger lava flows in the basin floor.
- Restoring the elevation lost by using a subsidence rate of 0.8 inches per year, Haleakala's ancient summit had an elevation of 13,500 and 14,500 feet.

- Source: USGS website: hov.usgs/volcanowatch/archive/2000/00_02_10.html.

MAUI CAPTURES THE SUN

- One cannot visit or discuss Haleakala without discussing the Hawaiian God Maui (sometimes called a demi god).
- The legend of Maui is known on many Polynesian Islands.
- His most famous feat is the *Legend of Haleakala and His Capturing the Sun.*
- Maui and his mother, Hina, lived near Rainbow Falls in Hilo on the Island of Hawaii.
- Hina would make kapa (a type of cloth) from the bark of the *wauke* and *mamaki* trees.
- By day, she would dye strips of kapa, but when night came, the kapa was still damp.
- She complained to Maui that the sun moved too quickly across the sky to dry the cloth.
- So, Maui traveled to Maui and climbed to the 10,023-foot summit of Haleakala where the sun was asleep in the crater.
- Maui, hiding in the crater until morning, watched the sun begin its daily journey.
- As the first ray of sunshine appeared, Maui snared the sun with his lasso of twisted coconut fiber.
- The sun demanded to be released, but Maui would not let go.
- Maui said to the sun: "Promise me you will move more slowly across the sky".
- Having no choice, the sun made a bargain with Maui.
- The sun said he would move more slowly for six months out of the year, and then he would move at his preferred pace for the other six months.
- Agreeing, Maui hurried home and told his mother the good news.
- As a reward, Hina made Maui a new cape that dried in one afternoon.

- For more information go to : **https://www.tourmaui.com/maui-demigod/**

- **Viewing the sun rise over Haleakala is a very popular park activity.**
- **However, reservations to view the sunrise are needed. see page 7.**

- **To see a 19 second time lapse video of the sun rising over Haleakala go to:**

http://www.aloha-hawaii.com/maui/haleakala/

END OF THE ROAD LOG

....

42

ROAD LOG FOR HANA TO THE KIPAHULU SECTION OF HALEAKALA NATIONAL PARK
(ROAD TO THE "SEVEN SACRED POOLS")

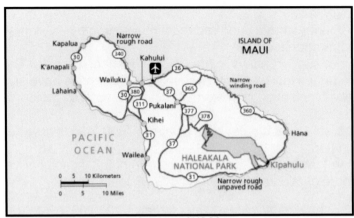

Courtesy of USGS

Preview: **Total Distance: 9.6 miles**

- Driving south from Hana you will be on Route 31 (also called Route 360).
- South of Hana you will pass some Hana age cinder cones located on Haleakala's east rift zone (described as   the least active of Haleakala's three rift   zones). *See the map below:*

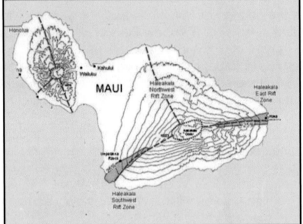

Modified from: http://files.hawaii.gov/dbedt/op/gis/maps/geothermal_maps.pdf

- The lava flows on this route are Holocene and Pleistocene Hana Volcanics and Pleistocene Kula Volcanics.
- As you proceed towards the so-called *"Seven Sacred Falls"* in *the Kipahulu area of the southern portion* of Haleakala National Park, you will pass Koki Beach, Hamoa Beach, Paihi Falls, and Wailua Falls. These places have been included in this road log.
- **Warning:** After leaving the Kipahulu section of Haleakala National Park, you may be temped to drive back to your hotel via the road to Kaupo and beyond.
- *Your rental car contract maybe voided if you use this road on the south side of Haleakala.*
- According to island guide books, the rental car agencies consider you to be on your own if you have any mechanical problems on this road.
- *However, if you use this route, **do not** attempt to complete this drive if you cannot finish it during daylight hours, especially in the unpaved portions.*
- ***Do not** drive this road in the dark. You may be all alone.*
- You will be driving along the south side of Haleakala, and the road will be winding, narrow, and in places unpaved. You may not have any cell phone service.
- **This author takes no responsibility for your actions.** • *Continued on the next page.*

- On the map below, the town of *Hana* is located at the *far* eastern end of Maui (in the green Hana lava area). The *Kipahulu area* is located in the same light green color west of Hana.

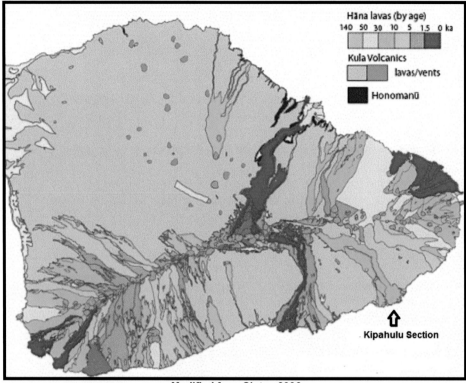

Modified from Sinton 2006

ROAD LOG FROM HANA TO THE KIPAHULA AREA OF HALEAKALA NATIONAL PARK

SET YOUR ODOMETER TO 0.0 MILES AT THE SERVICE STATION JUST PAST THE HASEGAWA GENERAL STORE.

0.0 miles: The service station just past the Hasegawa General Store.
- Some maps label the road south of Hana as Route 360, other maps label it Route 31.
- But, use caution because this road becomes very winding and narrow.
- **Also, in this direction, the milepost numbers are decreasing.**

0.1 miles: Sign Haleakala National Park: 9.5 miles
.
0.3 miles: Looking inland, if the vegetation is not too tall, the East Rift Zone *Puu Kolo* cinder cone can be seen. *First photo to the right.*
- Also, you are driving though old sugar cane fields.

0.9 miles: On the left, the *Kaiwiopele* cinder cone, another East Rift Zone volcano, is visible. *Second photo to the right.*
- According to a Hawaiian legend, the Volcano Goddess *Pele* originally came from Tahiti. But, she fled to the Hawaiian Islands to escape the wrath of her older sister *Namaka*, whose husband *Pele* had seduced!
- But, *Namaka* was the Goddess of the Sea, and because water is more powerful than fire, *Pele* was killed, and her body was torn apart.
- The name *Kaiwiopele* means *the bones of Pele*. • *Continued on the next page.*

- But, this was not the end of the story. After Pele died, her *spirit* flew to the Island of Hawaii, finding a home on the Kilauea Volcano. Pele seems to have 9 lives and really got around!

1.2 miles: On the right, a number of East Rift Zone cinder cones are visible,

1.4 miles: Haneoo Road.
- **TURN LEFT** onto Haneoo Road to access Koki Beach and Hamoa Beach.
- At the intersection of Route 31 and Haneoo Road, the above mentioned red cinder cone, *Kaiwiopele,* will be on your left.
- Drive along the shoreline to Koki Beach and Hamoa Beach.

1.8 miles: Koki Beach, meaning *the very top, as in culmination.*
- This is the first beach seen as you proceed down this road.
- The *Kaiwiopele* cinder cone is on the left side of this beach. Its western wall was eroded by wave action and flooded by the sea to form this popular beach. *First photo to the right.*
- The sand here is a mixture of white coral and black sand eroded from the adjacent rocks.
- During the summer months, a wide, flat, sandy beach is present. But, during the winter and spring, heavy surf erodes this beach back to a narrow strip of sand at the water's edge.
- Due to the presence of offshore boulders and rip currents, swimming here is not recommended

- *Also, walking close to this cinder cone is hazardous because volcanic debris can fall to the beach.*
- On the left, in the *far distance*, the *Lehoula* sea arch might be visible. *Second photo to the right.*
- Seaward, another eroded cone formed *Alau Island,* which today, is a seabird sanctuary. *Third photo to the right.*
- This island, topped by two coconut trees, was once a sacred Hawaiian site.

- *According to one website,* the two coconut trees were planted to commemorate a father losing his sons in World War II.
- However, a Hawaiian legend says *Alau* is a lost remnant of Maui left behind when the gods pulled the islands from the sea to create the Hawaiian Island chain.
- Also, because of its orientation, Koki Beach accumulates lots of debris washed onshore.
- Lastly, this entire coastline is at sea level, and thus, it is in a dangerous tsunami wave zone.

2.4 miles: Hamoa Beach:
- Access to this beach is via a path next to the E9 telephone pole, or by walking down a service road at about 2.6 miles. *Do not drive down this service road!*
- Because of its limited entry, it is easy to drive past this beach.
- Hamoa Beach is one of the most beautiful beaches on the Hana side of Maui. *Photo to the right.*
- Hamoa Beach is consistently named one of Maui's Best Beaches.
- Surrounded by sea cliffs, Hamoa Beach is approximately 1,000 feet long and about 100 feet wide.
- Hamoa Beach is described as the best body surfing beach on the island.
- *Do not try bodysurfing during high waves! These waves can pound you into the bottom!*
- Hamoa Beach is exposed to the open ocean, and thus, powerful currents and surf can often be present. Use great caution if you enter the water.
- *Continued on the next page.*

- Hamoa Beach is used mainly by the guests of the Hotel Hana Maui.
- Since the 1930's, the hotel has maintained this beach and its amenities for their guests.
- When the Hotel Hana Ranch (now named the Hotel Hana Maui) was being constructed, this beach was called Hamoa Beach. Prior to this time, it was called Mokae.
- Incidentally, Oprah Winfrey bought the *Ka Iwi O Pele* cinder cone, the old *Mokae Point Landing,* and several hundred acres of nearby land on which she has built a home.
- However, she donated 182 acres of her property to the Maui Coastal Land Trust to conserve and maintain it as open land.
- Continue to drive down Haneoo Road

2.6 miles: On the left, a tsunami warning siren.
- Again, being this close to sea level, this warning siren is critical to the people living here.
- Continue down this road until you reach the intersection of Haneoo Road and Route 360 at 2.9 miles. **Then do the following:**

RESET YOUR ODOMETER TO 0.0 MILES AT THE INTERSETION OF HANEOO ROAD AND ROUTE 31 AND TURN LEFT.

0.0 MILES: INTERSECTION OF HANEOO ROAD AND ROUTE 31.
- **At this intersection, turn left.**
- *Warning, this road becomes very narrow and bumpy.*

2.0 miles: You have been on Hana lava flows since you left Hana, and you will be on Hana lava for approximately 5-6 miles south of the town of Hana.

2.6 miles: At the curve in the road, a blocky basalt is underlain by a red laterite soil.
- This arrangement indicates that an older lava flow or sediments were exposed on the surface for a significant time.
- Consequently, this parent material was chemically weathered to form this red lateritic soil.
- The soil's red color is due to the iron bearing minerals in the parent material being oxidized and the resulting iron oxide staining the surrounding soil grains red.

3.1 miles: On the left, a Green Bay Packer sign.
First photo to the right.

3.7-3.8 miles: On the left, through the trees, and up on the hilltop, a **white cross** was erected in the memory of Father Helio Koaeloa, one of the earliest Catholic priests on Maui. *Second photo to the right.*
- Father Koaeloa was born in 1815 and died in 1846.
- This cross was erected in his memory in 1931.

3.9 miles: Paihi Falls. *Third photo to the right.*
- This roadside waterfall has a 50 foot drop.
- However, the flow of this fall is greatly dependent on the amount of recent rainfall.
- Parking is limited.

46

4.2 miles: Wailua Falls:
- *After you pass the falls, parking is on the ocean side of the road.*
- This stream drops 95 feet as it flows over vertical columnar basalts. *First photo to the right.*
- Wailua Falls is described as one of the prettiest waterfalls in Hawaii.
- This waterfall is located on the Honolewa Stream and not the Wailua stream.
- The Honolewa Stream passes under a bridge on the road.

4.8 miles: Hana lavas.

6.0 miles: SIGN: Haleakala National Park, Kipahulu Area.
See the map below.

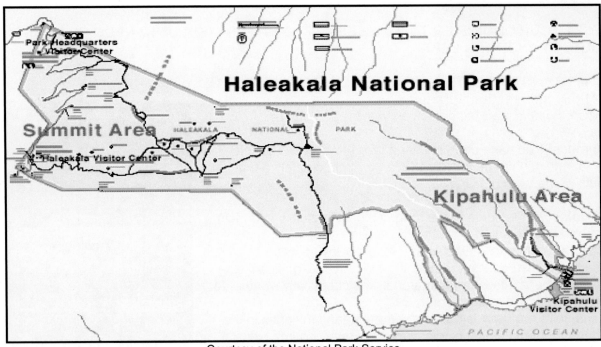

Courtesy of the National Park Service

- The above map shows that Haleakala National Park consists of two sections: The **Summit Area** and the **Kipahulu Area.**
- *There is no road that joins these two sections together.*
- The only access to the Kipahulu Area is by driving from Hana, or by the less recommended and hazardous road that goes along the south side of Haleakala (Route 31, which is unpaved in some sections).
- **For a good description of the Kipahulu Haleakala National Park Area go to:**

  https://www.nps.gov/hale/planyourvisit/kipahulu.htm

6.4 miles: A shrine to the Virgin Mary is located in a lava tube
Second photo to the right.
- This shrine was built by a local priest.

6.7 miles: Sign: Haleakala National Park *Kipahulu Area*. *Third photo to the right.*
- The **Oheo Bridge** is also at this mileage point. *Photo to the lower right.*
- As you cross the bridge, looking to the right, you can see the Oheo Gulch, and the *Seven Sacred Pools. Lower left and middle photos below.*
- Some sources say that this series of waterfalls and pools is rated as *the* most popular attraction in east Maui.

Swimming at the Pools,
- After an accident and lawsuit in 2009, jumping into the pools and swimming in them is tightly controlled by the National Park Service. Worse yet, on April 14, 2013, a 52-year-old Maui man died after falling 12-feet into a pool on the ocean side of the highway.

- *According to the park's website:*
   "Swimming is possible in the Kipahulu District of the park, but only when conditions allow. The freshwater pools at Oheʻo Gulch are prone to very dangerous flash floods caused by heavy rains high on the mountain. Injuries and deaths have occurred. Always check at the visitor center for current conditions before entering the water, obey all posted signs, and use your best judgment. *Diving and jumping are prohibited."* Do not bring glass into the Oheo Gulch area. Due to rough conditions, there is no safe ocean entry in the Kipahulu"

- **Note:** As stated above, you should be extremely cautious in this area; due to rapidly changing weather conditions, flash flooding can occur with little or no warning. If you hear a loud rumbling get away from the water and get quickly to high ground.
- To preserve this area, Laurence Rockefeller, The Nature Conservancy, and others, purchased 52 acres around the pools and donated this land to the Haleakala National Park.

The Naming of the Seven Sacred Pools

- The name, *The Seven Sacred Pools*, began many years ago when the owner of the Hotel Hana Maui wanted to attract people to his hotel.
- He didn't think calling this area *Oheo Gulch* generated much excitement.
- So, he called these pools The Seven Sacred Pools; which are not sacred and do not number seven.
- *But*, his advertising worked! *You are here!*

7.0 miles: On the left, the entrance to Hawaii National Park Kipahulu Parking Area. *Photo to the lower left.* Pay fee if applicable.
● A visitor center is located past the restrooms. *Middle photo below.*
● Outstanding scenic sights are along the following two hiking trails. *Photo to the lower right.*

  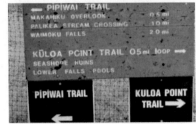

● The **Kuloa Point Trail** goes down toward the shoreline with views of the Seven Sacred Pools. *See the map below.*
● However, the most famous hiking trail is the **Pipiwai Trail**. This 4 mile round trip hike, that gains 650 feet in elevation is located above the Oheo Gulch.
● 2.5 to 5 hours are required to complete this hike. Bring water, a hat, and sun screen.
● One-half mile up the trail you will come to Makahiku Falls (with a 185 feet drop). *Photo to the lower left.*
● Then, you will come to a bamboo forest. *Middle photo below.*
● Another 1.5 miles on the trail leads to the base of *Waimoku Falls* (with a 450 feet drop). *Photo to the lower right.*

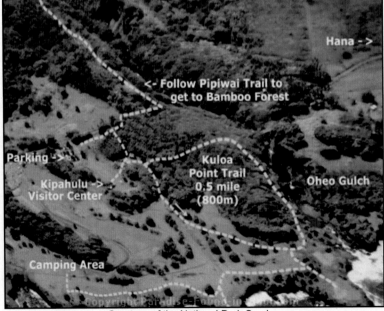

Courtesy of the National Park Service

Prior to leaving the park's parking lot, you might want to read the following material regarding the Kipahulu Valley and the Kipahulu Biological Reserve.

THE HALEAKALA BASIN, THE KAUPO GAP, AND THE KIPAHULU VALLEY

- The **Kaupo Valley** is the large south facing valley in the center of the image below.
- The **Kaupo Gap,** *cutting through the rim of the Haleakala Basin,* is at the head of the Kaupo Valley.
- See pages 32 and 36 for a discussion of the Kaupo Gap.
- **The Kipahulu Valley** *is located to the right of the Kaupo Valley* on the image below.
- It extends from sea level all the way up to Haleakala's rim. *But this valley it does not break through the rim. See the drawing below.*

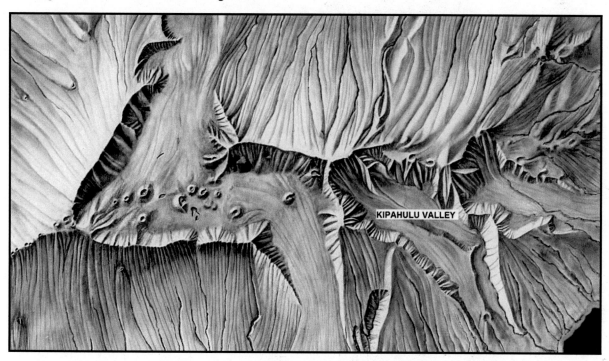

Courtesy of the National Park Service

- In the above image, the Kipahulu Valley is wide at the top and narrows downward forming a delta at the shoreline as the Palikea Stream meets the sea.
- From Haleakala's summit, Hana lavas (Maui's youngest lavas) flowed down this valley to the sea.
- Today, buried by sediments, these lavas are beneath the valley floor. However, they are exposed in the Palikea Stream Gulch (Hazlett and Hyndman, 1996)
- Also, in some places, the Hana lavas, after being exposed on the surface for a considerable period of time, were chemically weathered to form red lateritic soils.
- Later, additional Hana lavas buried these soils.

KIPAHULU VALLEY DETAILS

- In 1786, when the first Europeans arrived, the French explorer, Captain La Pérouse noted that because this land was productive, thousands of Hawaiians lived in Hana, Kipahulu, and Kaupo.
- Consequently, Kipahulu is rich in cultural features, such as, the remains of old agricultural terraces, shelters, fishing shrines, temples, canoe ramps, and rock walls.
- The Kipahulu coastal villages consisted of single story, one room buildings with thatched walls and roofs. ● *Continued on the next page.*

50

- These buildings acted as living quarters, cook houses, work sheds, storage, and canoe huts.
- However, in the 1880's, when the whaling industry came to Maui, the native people moved to places like Lahaina, causing a decline in the Kipahulu population.

THE KIPAHULU VALLEY BIOLOGICAL RESERVE

- The *upper* **Kipahulu Valley** is one of the most remote valleys in Hawaii. It is also the site of one the most pristine forests in all of Hawaii. *See the map below.*

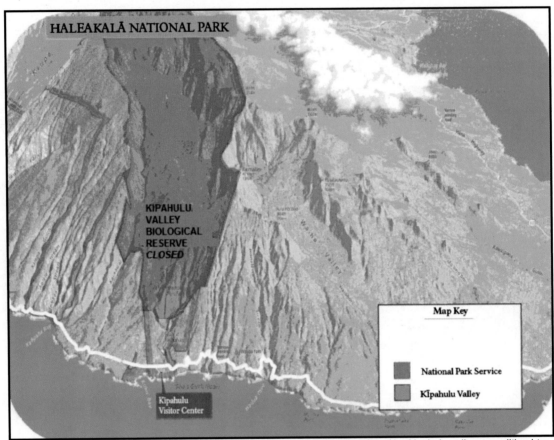

Modified from National Park website https://www.nps.gov/hale/learn/historyculture/the-kipahulu-valley-expedition.htm

- Consequently, the endemic birds and plants in this valley are found nowhere else in the world.
- To prevent devastating non-native species from entering this area, no trails or roads are planned for this area
- Entry into the Kipahulu Valley is only given to scientists conducting research and management studies essential to protecting this rare ecosystem.
- In 1969, due to the efforts of Charles Lindbergh, Sam Pryor (a long time Kipahulu resident), Laurence Rockefeller, and The Nature Conservancy; the Kipahulu Valley (from an elevation of 10,000+ feet down to sea level) was added to Haleakala National Park.
- However, this unique area is currently endangered by encroaching invasive plant species like Koster's curse, kahili ginger, and strawberry guava.
- These rapidly spreading plants out compete native rainforest plants, and they destroy critical habitats needed for native birds.
- Fences now protect the upper valley against the ravages of goats and pigs.
- For additional information on the 1967 scientific expedition into the Kipahulu Valley go to:

https://www.nps.gov/hale/learn/historyculture/upload/Kipahulu-Valley-Expedition.pdf

END OF THIS ROAD LOG

GLOSSARY OF SELECTED VOLCANIC TERMS

Modified from: USGS Photo Glossary of volcanic terms.htm

Aa or blocky lava (pronounced "ah-ah"): a Hawaiian term for lava flows with a rough rubble-covered surface composed of broken lava blocks called clinkers. The irregular surface of a blocky flow is very difficult to walk over.

Ash: consists of rock, mineral, and volcanic glass fragments smaller than 2 mm (0.1 inch) in diameter, but particles less than 0.025 mm (1/1,000th of an inch) in diameter are common.

Basalt: a hard, black volcanic rock that formed from a cooling lava. Some basaltic lava flows are very fluid and they can quickly flow more than 12 miles from a vent. Basalts are erupted at temperatures between 1100 to 1250° C. Most of the ocean floor is made of basalt, and the shield volcanoes of the Island of Hawaii are composed almost entirely of basalt.

Blocks: solid rock fragment greater than 64 mm (2.52 inches) in diameter ejected from a volcano during an explosive eruption. Blocks commonly consist of solidified pieces of old lava flows that were part of a volcano's cone.

Bomb: droplets of lava ejected from a volcano while they were partially molten.

Calderas: a large, usually circular, depression at the summit of a volcano. Calderas form when large volumes of magma are withdrawn into the subsurface or erupted from side vents. With the structural support for the overlying rock removed, the walls of the volcano slump inward to form this large summit opening.

Craters: circular summit depressions much smaller than that of a caldera.

Cinders: partly vesiculated (that is, contains gas bubble holes) basaltic lava fragment ejected during an explosive volcanic eruption.

Cinder cones: steep, conical hills, generally less than 1500 feet high, composed of volcanic fragments that accumulated around a volcanic vent

Dikes: tabular rock bodies that cross cut the layering of the adjacent rocks.

Ejecta: a general term for rock particles ejected into the air by a volcano.

Fissures: elongate fractures or cracks from which lava erupts.

Gas: Magma and lavas contain dissolved gases that are released into the atmosphere during a volcanic eruption. The most common gas released from a volcano is steam (H_2O), followed by CO_2 (carbon dioxide), SO_2 (sulfur dioxide), (HCl) hydrogen chloride, and other compounds.

Lava: molten rock erupted onto the Earth's surface.

Lava flows: masses of molten rock that poured onto the Earth's surface during a non-explosive volcanic eruption.

Lava fountains: form when jets of lava are sprayed into the air by the rapid formation and expansion of gas bubbles in the molten rock. Lava fountains typically range from 30 to 300 feet in height, but occasionally they reach more than 1600 feet high.

Lava tubes: natural conduits through which lava travels beneath the surface of a lava flow. Tubes form by the crusting over of lava channels and pahoehoe (ropy) flows. At the end of an eruption, the lava in the tube drains away to form hollow lava tubes.

Lava velocities: The fastest recorded Hawaiian lava flow was in the 1950 Mauna Loa southwest rift zone eruption. The front of this flow advanced from its vent to Highway 11 at an average speed of 6 miles/hour. By contrast, a typical Big Island Puu Oo blocky flow moves at less than 1/3 of a mile/hr. Lava can flow faster than 6 mi/hr if the lava is in a lava tube or channel. There, because the lava is confined, it can stay hot. Lava velocities within a lava channel have been measured at nearly 35 mi/hr during the 1984 Mauna Loa eruption. During the first years of the Puu Oo eruptions, lavas within lava tubes had speeds up to 23 miles/hr.

Magma: molten rock beneath the Earth's surface. Conversely, when magma erupts onto the surface, it is called lava.

Minerals: naturally occurring substances that are inorganic in composition, have an orderly internal arrangement of their atoms or ions, have a definite or nearly definite chemical composition, and have definite physical properties, such as density and hardness. In contrast, *Rocks* are composed of one or more minerals

Pahoehoe (or ropy lava): a Hawaiian term for basaltic lava flows that have a ropy surface. As a fluid pahoehoe lava flow cools, a thin "skin" forms on the surface of the lava. Then, due to the movement of the lava below, "wrinkles" on the overlying cooled surface develop into a ropy pattern.

Pumice: a gray, glassy, frothy igneous rock that forms during explosive eruptions, or when the top of a lava flow is "whipped" into a glassy foam by escaping gases. Because of its foamy texture, it has a low density and thus, it will float on water. Also, pumice is commonly used as an abrasive stone.

Pyroclastic debris: fragments produced by a violent volcanic eruption (*pyro* meaning fiery, and *clastic* meaning broken). Pyroclastic debris includes ash, cinders, block, and bombs.

Pyroclastic flows: an avalanche of hot ash, pumice, rock fragments, and volcanic gas that rushes down the side of a volcano at speeds up to 60 or more miles/hour. The temperature within a pyroclastic flow may be greater than 1000° F.

Rift zones: fractures in the Earth's crust that form when the crust is pulled apart. Hawaiian shield volcanoes generally have three rift zones; along which cinder cones, spatter cones, pit craters, lava flows, lava fountains, ground cracks, and normal faults can form.

Rocks: a naturally occurring assemblage of minerals. However, there are some one mineral rocks like limestone, chert, and dunite.

Shield volcanoes: big, broad volcanoes with gentle slopes formed by the eruption of fluid basaltic lava. These volcanoes commonly have flank eruptions like Kilauea's Mauna Ulu or Puu O oo. The largest volcanoes on Earth are shield volcanoes. This name comes from the volcanoes' resemblance to the shape of a warrior's shield.

Vesicular: gas bubble holes in a basalt. If these holes are fill-in with a secondary mineral, it is called an amygdaloidal basalt.

Volcanoes: conical hills or mountains that form when lavas, gases, and pyroclastic debris is extruded from a central vent.

   Active volcanoes: have erupted in historic times.

   Dormant volcanoes: volcanoes that have not erupted in history times, but shows no erosion of the cone, but they could erupt again.

   Extinct volcanoes: have not erupted in historic time and show erosion of the cone. Thus, extinct volcanoes are considered unlikely to erupt again. Whether a volcano is truly extinct is often difficult to determine. *However, other sources might give a different definition of "extinct".*

Volcanic gases: dissolved gases that are released into the air when lava is extruded onto the surface. Gases can also be released from subsurface magma through volcanic vents, fumaroles, and hydrothermal systems. The most common gas released by magma is steam (H_2O), then CO_2 (carbon dioxide), SO_2 (sulfur dioxide), (HCl) hydrogen chloride, and other compounds.

SIMPLIFIED GEOLOGIC TIME SCALE

Maui's lavas Holocene, Pleistocene and Pliocene in age. Thus, the time scale below has been enlarged to just show the subdivisions (epochs) of the Tertiary and Quaternary Periods.

Ages of the units shown in the table below are in millions of years.

This time scale shows that Maui's Holocene, Pleistocene, and Pliocene lava flows are, geologically, very young.

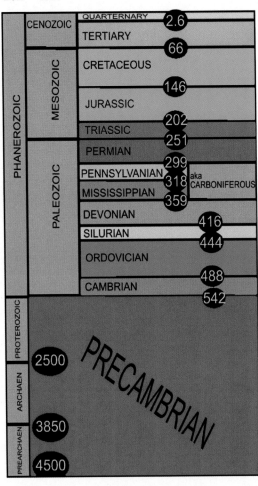

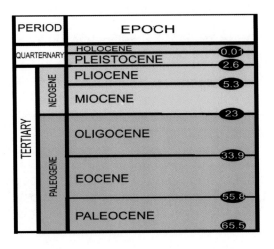

THE EARTH IS ABOUT 4.6 BILLION YEARS OLD

STRATIGRAPHIC AGES OF MAUI LAVA FLOWS

Hana Volcanics: Qhn6: 0-1500 years ago to Qhn0: 50,000 to 140,000 years ago

Kula Volcanics: 0.93 to 0.15 million years ago

Honomanu Basalt: 1.1 to 0.97 million years ago.

Lahaina Volcanics: 0.6 to 0.3 million years ago

Honolua Volcanics atop Wailuku: 1.3 to 1.1 million years ago.

Wailuku Basalt: 2.0 to 1.3 million years ago.

From: **Haleakalā National Park,** *Geologic Resources Inventory Report* Natural Report NPS/NRSS/GRD/NRR—2011/453

SYMBOLS USED ON MAUI GEOLOGIC MAPS FOR *SEDIMENTS*

Qa - Alluvium (Holocene)

Qbd - Beach deposits (Holocene)

Ody Younger dune deposits (Holocene)

Odo Older beach deposits (Holocene and Pleistocene)

QTao - Older alluvium (Pleistocene and Pliocene)

SYMBOLS USED ON MAUI GEOLOGIC MAPS FOR *HALEAKALA VOLCANICS*

Qhn- Hana Volcanics (Holocene and Pleistocene)

Qkul - Kula Volcanics; lava flows (Pleisticene)

SYMBOLS USED ON MAUI GEOLOGIC MAPS FOR THE *WEST MAUI VOLCANO*

Qlhl Lahaina Volcanics (Pleistocene)

Qul Honolua Volcanics (Pleistocene)

Qtwl Wailuku Basalt (Pleistocene and Pliocene?)

MAUI GEOLOGIC MAPS

The following geologic maps were taken from selected portions of the:

Geologic Map of th State of Hawaii, Sheet 7 -Island of Maui
by
Sherrod, D.R, Sinton, J.M, Watkins, S.E. and Brunt, K.M., 2007

NOTE: These maps have been saved as jpeg files. *To make some of the maps more readable,* the jpeg file have been enlarged or stretched. Thus, parts of the map have been distorted for easier readability of the formation names.

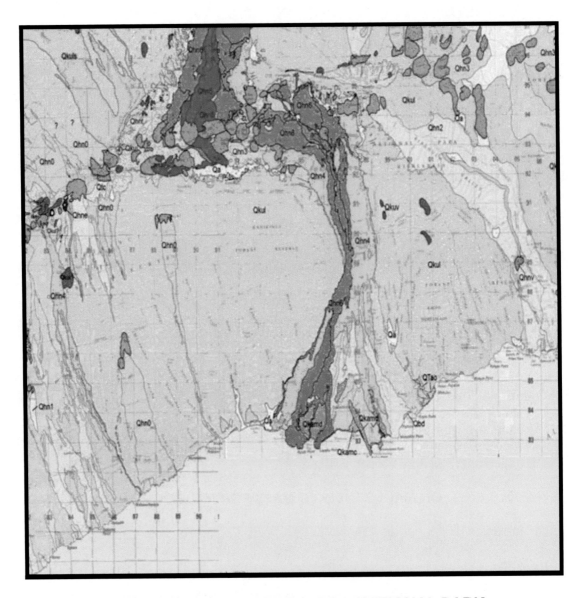

**GEOLOGIC MAP OF HALEAKALA NATIONAL PARK
FROM SUMMIT TO SHORELINE**

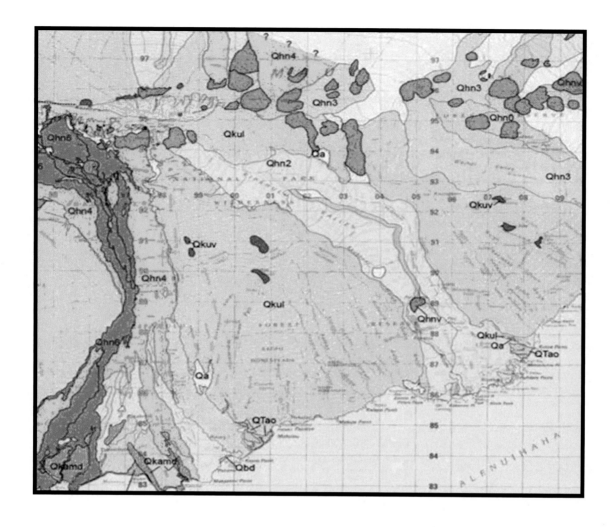

GEOLOGIC MAP OF HALEAKALA NATIONAL PARK KIPAHULU SECTION

REFERENCES USED IN THIS BOOK

Decker, Robert, and Decker, Barbara, Road Guide to Haleakala and the Hana Highway, 1992.

Doughty, A., Maui Revealed, 2012

Hazlett, R. W., Hyndman, D. W., Road Side Geology of Hawaii, 1996.

Kirch P. V., Hartshor A. S., Chadwic O. A., Vitouse P. M., Sherrod D. R., Coil J., Holm L., and.. Sharp, D. W., 2004, Environment, agriculture, and settlement patterns in a marginal Polynesian Landscape. Kyselka, W. and Lanterman, R, Maui How It Came To Be, 1980.

Porter, S. C., M. Stuiver, and I. C. Yang. 1977. Chronology of Hawaiian Glaciations. Science 195(4273):61–62.

Sinton, John, Maui Field Guide, 2006.

Thornberry-Ehrlich, T. 2011. Haleakalā National Park: geologic resources inventory report. Natural Resource Report NPS/ NRSS/GRD/NRR—2011/453. National Park Service, Ft. Collins, Colorado.

Haleakalā National Park Geologic Resources Inventory Report Natural Resource Report NPS/NRSS/GRD/NRR—2011/453

Natural History of the HawaiianIslands, edited by E. Alison Hay 1994

https://en.wikipedia.org/wiki/Hosmer%27s_Grove

http://www.gohawaii.com/maui/regions-neighborhoods/upcountry-maui/kula

http://www.to-hawaii.com/maui/cities/kula.php

http://hawaiitrails.ehawaii.gov/trail.php?TrailID=MA+18+001

http://pubs.usgs.gov/sir/2012/5010/sir2012-5010.pdf

http://hvo.wr.usgs.gov/volcanowatch/archive/2002/02_06_27.html

http://myhawaii.biz/maui-geological-history/

http://www.gohawaii.com/maui/about/history

www.gohawaii.com/maui/...

http://hawaiitrails.ehawaii.gov/trail.php?TrailID=MA+18+001

http://pubs.usgs.gov/sir/2012/5010/sir2012-5010.pdf

http://hvo.wr.usgs.gov/volcanowatch/archive/2002/02_06_27.html

http://myhawaii.biz/maui-geological-history/

http://www.gohawaii.com/maui/about/history

http://bulletin.geoscienceworld.org/content/115/6/683

http://www.hawaii-guide.com/maui/articles/maui_weather

http://hvo.wr.usgs.gov/volcanowatch/archive/2001/01_09_13.html

http://www2.hawaii.edu/~nasir/0811MauiGeology.ppt#257,1,Slide 1

http://volcano.oregonstate.edu/vwdocs/volc_images/north_america/hawaii/Kahoolawe_map.gif

http://www.haleakala.national-park.com/info.htm

http://www.pnas.org/content/101/26/9936.full.pdf

http://www.hawaiilife.com/articles/2012/01/hawaiis-winds/

http://www.world-guides.com/north-america/usa/hawaii/maui/maui_weather.html

http://www.sciencedirect.com/science/article/pii/S0377027302003852

http://www.unrealhawaii.com/wp-content/uploads/2013/08/Map-and-Descriptions.pdf

http://hvo.wr.usgs.gov/volcanowatch/archive/2003/03_04_10.html

http://www.hawaii-guide.com/maui/sights/oheo_gulch_kipahulu

http://online.wsj.com/news/articles/SB10001424052748704320104575015431045157378

http://hvo.wr.usgs.gov/volcanowatch/archive/2001/01_02_08.html

http://www.fourwindsmaui.com/molokini-crater

http://tourmaui.com/maui-ancient-history/

http://pacificscience.files.wordpress.com/2014/02/pac-sci-early-view-68-3-4.pdf

http://www.gohawaii.com/maui/regions-neighborhoods/upcountry-maui/kula

http://www.to-hawaii.com/maui/cities/kula.php

http://www.fodors.com/world/north-america/usa/hawaii/maui/review-414853.html

http://hawaiitrails.ehawaii.gov/trail.php?TrailID=MA+18+001

the above source has rain fall map

http://pubs.usgs.gov/sir/2012/5010/sir2012-5010.pdf

http://hvo.wr.usgs.gov/volcanowatch/archive/2002/02_06_27.html

http://myhawaii.biz/maui-geological-history/

http://www.gohawaii.com/maui/about/history

http://bulletin.geoscienceworld.org/content/115/6/683

http://www.hawaii-guide.com/maui/articles/maui_weather

http://hvo.wr.usgs.gov/volcanowatch/archive/2001/01_09_13.html

http://www2.hawaii.edu/~nasir/0811MauiGeology.ppt#257,1,Slide 1

http://volcano.oregonstate.edu/vwdocs/volc_images/north_america/hawaii/Kahoolawe_map.gif

https://mauiguide.com/wp-content/uploads/edd/2017/09/maui-guide-ebook-1.pdf

www.hawaiistateparks.org/parks/maui/index.cfm?park_id=36

www.hawaiiweb.com

www.hawaiistateparks.org/parks/maui/index.cfm?park_id=36

www.hawaiiweb.com

....

Made in the USA
Las Vegas, NV
27 December 2022

64252280R00036